Black Powder Revolvers — Reproductions & Replicas
By Dennis Adler

Publisher's Suggested Retail Price - $39.95

Buffalo Bill's Wild West Tribute Revolver

To commemorate the dual anniversaries of the Remington Model 1858 revolver introduced 150 years ago and the 90th Anniversary in 2007 of the last Buffalo Bill's Wild West Show, America Remembers introduced the Buffalo Bill's Wild West Tribute Remington New Model 1858 revolver in a limited edition of 300 guns. Designed by the author, America Remembers Chairman Paul Warden and the craftsmen of A&A Engraving, the revolver features beautiful 24-karat gold etched images of Col. Wm. F. Cody on the cylinder, his signature on the backstrap, decorative scrollwork, and the barrel inscription, "1883 - *Buffalo Bill's Wild West* - 1917". The presentation pistol comes with a glass top display case and letter of authenticity. Of his Remington revolver, Col. Cody proudly stated, "It never failed me."

Publisher's Note:

No part of this publication may be reproduced in any form whatsoever, by photograph, mimeograph, fax transmission or any other mechanical or electronic means. Nor can it be broadcast or transmitted, by translation into any language, nor by electronic recording or otherwise, without the express written permission from the publisher – except by a reviewer, who may quote brief passages for critical articles and/or reviews.

All Rights Reserved
Copyright 2008

ISBN 10: 1-886768-82-X
ISBN 13: 978-1-886768-82-6

Blue Book Publications, Inc.
8009 34th Avenue South, Suite 175
Minneapolis, MN 55425 U.S.A.
GPS Coordinates: N 44° 51 28.44, W 93° 13.1709
Orders Only: 800-877-4867, ext. 3 (domestic only)
Phone No.: 952-854-5229
Fax No.: 952-853-1486
General Email: support@bluebookinc.com
Web site: www.bluebookinc.com

Distributed in part to the book trade by Ingram Book Company and Baker & Taylor.

Distributed throughout Europe by Deutsches Waffen Journal
Rudolf-Diesel-Strasse 46
Blaufelden, D-74572 Germany
Website: www.dwj.de

Published in the United States of America, printed in South Korea

CREDITS:
Photography – Dennis Adler & S.P. Fjestad
Designer & Art Director – Clint H. Schmidt
Editorial – Dennis Adler, R.L. Wilson, & S.P. Fjestad
Proofreading – Sara Lange, Kelsey Fjestad, & Cassandra Faulkner
Printing – Pacom, South Korea

ABOUT THE FRONT COVER:

A historic evolution of black powder revolvers: a U.S. Historical Society c.1990 Jefferson Davis Commemorative; a 3rd Generation Colt Blackpowder Arms 1860 Army custom engraved and silver inlaid by Conrad Anderson; an 1858 New Model Remington Army hand engraved by Rocky Hayes and custom finished by Kirst & Strite; a handcrafted Colt 1860 Army Richards Type I cartridge conversion built and engraved by R. L. Millington; and an 1860 Army Richards-Mason snub nose conversion built and engraved by R. L. Millington. Cover photo by Dennis Adler.

ABOUT THE BACK COVER:

Clockwise from top left: The new 2008 Remington Arms Co. 150th anniversary 1858 New Model Army revolver; Kenny Howell-built 1851 Navy Colt conversion; cased and engraved pair of 3rd. Generation 1860 Army revolvers by John J. Adams, Sr.; hand engraved Pietta LeMat revolvers; pair of specially engraved Colt 1860 Army revolvers with cast ceramitation cloisonné embellishments and gold grips done by Andrew Bourbon for Tiffany & Co. and Dr. Joseph A. Murphy; custom-built Belt Model Paterson by R.L. Millington and Dan Chesiak; 1860 Army Richards Type I conversion by R.L. Millington, engraved and Tiffany-griped by John J. Adams, Sr.; antiqued Pietta Starr single and double action revolvers; and center, author Dennis Adler with a Starr single action cartridge conversion by R.L. Millington. Photos by Dennis Adler.

Contents

TITLE PAGE 1
COPYRIGHT, CREDITS, & COVER DESCRIPTION 2
TABLE OF CONTENTS 3
INTRODUCTION by S.P. Fjestad 4
FOREWORD by R. L. Wilson 5-7
PREFACE & ACKNOWLEDGEMENTS by Dennis Adler 8-9

CHAPTER ONE: Colt Percussion Pistols 10-37
From Patersons to Police and Navy Models

CHAPTER TWO: Colt's Contemporaries 38-59
Remington, Rogers & Spencer, Spiller & Burr, Starr, Dance, Griswold & Gunnison, and LeMat

CHAPTER THREE: Reproduction Percussion Pistols 60-73
In the Image of Col. Colt

CHAPTER FOUR: Limited Edition and Collectable Models 74-121
Engraved and Boxed For Presentation in the Colt Tradition

CHAPTER FIVE: Shooting And Maintaining Black Powder Pistols 122-141
Better Than James Butler Hickok Had It

CHAPTER SIX: Behind The Scenes 142-157
A Tour of The World's Leading Percussion Pistol Manufacturers

CHAPTER SEVEN: Practical Percussion Pistol Shooting 158-189
Skin That Smokewagon!

CHAPTER EIGHT: Metallic Cartridge Conversions 190-240
From Lead Balls to Brass Cartridges

TRADEMARK INDEX/PRODUCT AND SERVICE SOURCES 241-244
MUSEUMS OF INTEREST 245-246
INDEX 247-252

Introduction
By S.P. Fjestad

It was around closing time on Saturday afternoon approximately the third week in January of 1997 at the Las Vegas Antique Arms show being held at the Riviera Hotel. R.L. "Larry" Wilson came over to our crowded table and said, "You really need to meet a friend of mine. He's a very talented photojournalist, has worked mostly in the automotive industry, and would now like to do a book on guns if possible." Asking Larry if this could wait until the next morning, he quickly replied, "Dennis has been waiting most of the afternoon to meet you – I'll bring him over right now." After a few minutes, introductions had been exchanged, and we talked as we walked across the street from the convention ballroom to the hotel. It was very obvious from the start that Dennis Adler knew his automobiles extremely well, but I wasn't sure if he had enough horsepower to do a book on what he wanted, which was a publication on Colt blackpowder reproductions and replicas.

By the time the dealers starting packing up on Sunday afternoon, it had been decided that a book on black powder reproductions and replicas manufactured since 1959 in both Italy and the U.S. would be welcomed by the ever increasing amount of consumers who were purchasing black powder revolvers for shooting, reenactments, and collecting. Nothing had been done on this subject before, and it was about time, as there were already more reproductions and replica handguns in circulation than originals from the 19th century. During the next couple of months, it was agreed upon that a trip to Italy would be necessary to get all the information this new publication deserved. So in September, Dennis, Larry, and myself rendezvoused in Brescia in September, and this landmark trip included factory tours of the original Uberti, Pietta, Euroarms Italia, Armi Chiappa, and Beretta plants. Maybe the most memorable stop during our trip was a factory tour of the Ferrari plant in Maranello, where we discovered that many of the workers are gun enthusiasts!

Production of *Colt Blackpowder Reproductions & Replicas* began in earnest during the summer of 1998, and books were shipped from the printer in Hong Kong directly to the Colt Collectors Association host hotel in Denver for their debut. Walking out to the hotel's loading dock, I opened up a random box on top of the skid full of books. Upon fanning the pages of the first book I grabbed, Dennis & I were shocked to see an unbound, black and white image of a Japanese fisherman inserted between the pages! Our fears quickly went away after we examined the rest of the books, sans fisherman, and we were very happy with the quality and color reproduction. It clearly was a winner and six years later, it was sold out.

Fast forward to late 2007. Initially, we decided to print a slightly updated 2nd edition, but because so much has changed within the black powder industry in the past ten years, we decided to substantially improve and update the book, utilizing a landscape format in hardcover only. What you have in your hands is easily the best book ever published on black powder reproductions and replicas. Dennis Adler has once again raised the bar with his photojournalistic professionalism, providing you with a literary time machine capable of transporting you back to the turbulent days of America's Civil War and Old West, while also covering all the most recent black powder makes and models.

I would like to thank both Dennis Adler and R.L. Wilson for giving me their pep talk over ten years ago on the importance of black powder reproductions and replicas, a field which has exploded in popularity in recent years. I would also like to thank Clint H. Schmidt, our art director, for doing an exemplary job in the design and layout, utilizing this landscape format. Blue Book Publications, Inc. plans to follow this volume with a sister book entitled *Black Powder Long Arms - Reproductions & Replicas*. Also, don't forget about the *Blue Book of Modern Black Powder Arms* by John Allen, which provides detailed model descriptions, images, and up-to-date values on most discontinued and recently manufactured black powder arms – it's the perfect companion for this book. I feel privileged to have published the only books available on black powder reproductions and replicas, and would like to thank you for supporting these projects.

Sincerely,

S.P. Fjestad
Publisher
Blue Book Publications, Inc.

Luigi Chinetti, Jr. (c), whose dad won Le Mans overall in 1949, driving a Ferrari 166MM, the first win for the fabled marque at the famous French racetrack, Dennis Adler (r), author of *Ferrari – The Road From Maranello*, and S.P. Fjestad, outside the front entrance to the factory. Many Ferrari employees are also avid gun enthusiasts!

Authors only! Dennis Adler (l), S.P. Fjestad, and R.L. Wilson (r) are shown holding their individual tomes during the 1999 SHOT Show. While the books are long out of print, the authors are not!

Foreword
By R. L. Wilson

With the passing of William B. Ruger, Sr., in 2002 and Val Forgett, Sr., in 2005, the world of firearms and the shooting sports lost two of its superstars. The inscription on the backstrap of an engraved, presentation Ruger Old Army, serial no. 59, given to Val Forgett by Bill Ruger, sums it all up neatly –

"To Valmore Forgett, a valiant competitor, from his friend, William B. Ruger."

That gift revolver was from one giant in the firearms industry to another. The tribute to Bill Ruger, written by Val in *Ruger & His Guns*, could just as well have been a description of Val himself. The two men shared many qualities, and could discuss innumerable subjects as equals.

In the 19th century the giants of the arms industry were men such as Samuel Colt and Elisha King Root; Oliver Winchester and B. Tyler Henry; Horace Smith and Daniel B. Wesson; Eliphalet Remington and his sons; John M. Marlin; Christian Sharps; Henry Deringer; Eli Whitney; Simeon North; Samuel E. Robbins and Richard S. Lawrence; C.M. Spencer; and John M. Browning. For many enthusiasts of original Civil War era firearms, as well as modern day reproductions, Val Forgett and Bill Ruger were 20th century equals.

First and foremost, both men knew and loved guns, and both understood machines… not only how to design firearms but how they were made. Both kept abreast of trends in the business, and established their own original modes of operation, innovative products, and spheres of influence. A vital arrow in each of their quivers was a solid grounding in the history and traditions of the amazing, exciting, and potentially profitable field of gunmaking. Combining all of this, both men led fantastic lives and enjoyed every day. Furthermore, both realized that this industry and those who make it up, support it, and are customers of it have a clear vision of legacy – that what they did as hunters, shooters, sportsmen, industrialists, and as collectors would impact the future of America and the world.

We owe a great deal to the vision of Val Forgett, who more than any one person made the replica firearms industry what it is today. Val's efforts – valiant as Bill Ruger stated – plus the appeal of the original firearms and that of the 19th century gunmakers – got it all started back in the 1960s.

The Civil War Centennial, which began in 1961, combined with the celebration of anniversaries of that conflagration and other historical events, gradually led to a revival in antique guns and in the art of gunmaking. Developments in computer technology and the evolution of CNC machines fueled the revival of these beautiful 19th century firearms.

Paul McCarthy (c), Sr. Editor of Simon & Schuster, myself (r), and Bill Ruger, Sr., admiring a specially bound copy of the *Ruger & His Guns* book.

Beginning in the 1970s, legendary Colt blackpowder models were reborn at the hands of gifted craftsmen like Aldo Uberti (who built his first Colt 1851 Navy reproduction for Val Forgett in the late 1950s). With the grace and beauty of the originals, these Colt reproductions were released on the market and widely and enthusiastically received. By 1971, Colt was once again placing its historic name on the frames of guns which had not been manufactured since the mid-19th century. In one memorable instance, principally using a spectrograph, Colt's engineering department was able to build a superb replica of the Walker Model Colt in 1980 (ably assisted by Angelo Buffoli, then owner of Armi San Marco in Gardone, Italy).

The charm of these replicas (some even as miniatures), their strikingly handsome configurations, captivating histories, and their practicality all led to these fresh, re-introduced arms developing their own unique and ever-growing niche in the U.S. market. Here we are a decade after the groundbreaking publication of *Colt Blackpowder Reproductions & Replicas* in 1998, with an industry that has continued to grow and draw in fresh converts to the

Aldo Uberti (l), his daughter Maria Laura Uberti, and Wilson, at the Uberti factory, Gardone, Val Trompia, c. September 1997.

firearms world daily. The old-fashioned replica arms attract new buyers with the innate nostalgia, the muzzle-loading demand for careful precision, and the fact that today one can participate in so many exciting activities. Black powder reproductions are used in Cowboy Action Shooting events, historic Civil War re-enactments, and even in recreations of historic shootouts like the famous Northfield, Minnesota Bank Raid by the James and Younger Gang. Bob Boze Bell of *True West* magazine said that the "Defeat of Jesse James Days" raid is the most authentic re-enactment of any western event in the world.

That past and present so rightfully coexist is a testament to the men who built the American firearms industry of the 19th century. Despite relatively primitive communications, in the early days of mass production and parts interchangeability Colt, Remington, and others were blessed with the demands of an evolving, indeed, exploding United States market fueled by a national belief in Manifest Destiny, and with immigration at its peak. Landmark events like the Mexican War, the California Gold Rush, the Civil War, and the opening up of the American West played enormous roles. The successors to historic figures such as Sam Colt, Daniel B. Wesson, Oliver Winchester, William F. "Buffalo Bill" Cody, and Theodore Roosevelt – all of who were great lovers of firearms, hunting, and the shooting sports – have been men like Val Forgett, Sr., William B. Ruger, Sr., and Jr.; Colt's Board Chairman George A. Strichman; Winchester's Bill Talley; Remington's Mike Walker and C.K. Davis; Smith & Wesson's Carl Helfstrom; and Marlin's Frank Kenna – men who picked up an historic gauntlet and carried it forward. Bill Ruger once said, "There is no one who can design a firearm without reference, without some connection to an earlier design. Just as engines, fuels, and aerodynamics help define what an automobile will look like, the dynamics of a bullet and how it works, defines a firearm."

Visionaries have served the firearms industry in more ways than just design and innovation. The preservation of an American tradition has also fallen upon the shoulders of these leaders. There were those whose insight and leadership led to creating industry advocacy groups, like the National Shooting Sports Foundation, organizers of the annual SHOT (Shooting and Hunting Outdoor Trade Show). The same insightful leadership created the European equivalent of SHOT, known as IWA (International Waffen und Ammunition show), held annually in Nuremberg, Germany.

Leaders like Harlan Carter and Wayne LaPierre of the National Rifle Association revitalized that organization, including its shows and conventions. NRA Treasurer Wilson Phillips and investment banker and multi-faceted expert John R. Woods were influential in long-term financial planning, including Woods' inspired concept of the NRA Foundation Endowment programs – a vital guarantee of the organization's effectiveness into the future to protect our 2nd Amendment rights.

Legendary firearms are part of the very fiber of American history, something universally recognized by organizations such as the NRA, Boone & Crockett Club, the National Shooting Sports Foundation, and collectors of historic firearms from every era. Although antique arms collectors represent a relatively small percentage of the makeup of the firearms industry, a much higher number within collecting are persons of considerable wealth, power, and influence, which raises the profile of arms collectors and their ability to promote the field in positive ways at the social and/or business level. William B. Ruger and Valmore Forgett led the way by setting an example for so many others. What an honor to be among the relatively few alive today who knew these two giants in the throes of their extraordinary careers – and to have seen, in fact, experienced firsthand their impact and how that has led to the highly organized and powerful firearms and shooting sports industry we enjoy in modern times.

Because of the vision, dedication, and genius of these men and their colleagues, the future of guns, hunting, and the shooting sports has never been more promising. Consider only the segment of the industry that promotes activities involving replica firearms. These makers, dealers, distributors, and of course, consumers have made more of these historic arms today than did the original manufacturers! This accomplishment was spurred on by a veritable blizzard of arms books – also aided and abetted by television and movies – particularly westerns starring Clint Eastwood and Tom Selleck, both of who have paid great attention to the historic guns shown in their movies. In turn, their films have had an overwhelming influence on the world of historic reproductions. The History Channel has also played a memorable role in furthering the recognition and preservation of historic firearms. I can remember some 25 or so years back, PBS's Frontline produced the forward-thinking "Gunfight U.S.A." narrated by the late Jessica Savitch. My role in that production included an interview in which I discussed the founding fathers, who were often keen firearms enthusiasts, including George Washington and Thomas Jefferson. In the ensuing years, History Channel's American Experience broadcast the Riva Productions one-hour documentary, "Annie Oakley." Over 53 hours of firearms history, artistry, and adventure have been released by the History Channel for public consumption. Executives of Greystone Productions, producers of "Tales of the Gun" and "The Story of the Gun" have said that of all PBS programming, the firearms shows have received the highest demand and most positive reaction. Americans love history, thus it comes as no surprise that the Arms and Armor Galleries at the Metropolitan Museum of Art are one of the most popular of all the displays in that institution (the other of the two most in demand is the exhibition of mummies, in the Egyptian Galleries).

All these factors contribute to continuing a grand tradition established by our 19th and early 20th century forbearers. Quoting Val Forgett, Sr., in the First Edition of *Colt Blackpowder Reproductions and Replicas*: "In the beginning, it was [Aldo] Uberti for revolvers, Antonio Zoli made our first Zouaves and Mississippis, Pedersoli made the Kentucky rifles and pistols, and Luciano Amadi in the beginning acted as an agent to get all of this started. Antonio Zoli, Davide Pedersoli, and Aldo Uberti have all passed on, while their children continue their work and ensure the future of the industry." Now with Val's passing, his son Valmore III, whose experience in the firearms industry dates back to pre-teenage years, carries on a family tradition. With Beretta's acquisition of A. Uberti in 1999 (following Aldo Uberti's passing), the company can look forward to a promising future as part of the world's oldest firearms maker, while Ugo Gussalli Beretta can look to his sons Piero and Franco to carry on a centuries-old family tradition. There are also importers and retailers like the ladies at Taylor's & Co., the Kirkland family at Dixie Gun Works, and Mike Harvey of Cimarron F.A. Co., who not only import and sell reproductions of historic arms but also contribute to their design and creation, much in the way Val Forgett did in the 1960s and 1970s.

Dennis Adler's new revised and updated book on reproduction black powder revolvers and metallic cartridge conversions goes a long way to fanning the flames that ignite the passion held by every arms collector. Congratulations to Dennis and to Blue Book Publications, Inc. on their role in bringing antique arms back to the future – and good luck with this grand book that in itself adds immeasurably to the world's ever-growing fascination with fine, historic guns and the shooting sports.

R.L. Wilson

Val Forgett Sr. and Val Forgett III, in Paraguay, 1997, seeking arms for purchase and importation.

S.P. Fjestad (l), Franco Gussalli Beretta, and R.L. Wilson (r) at the opening of the new Beretta gallery at the National Firearms Museum, 2001.

Dr. Ugo Gussalli Beretta (c), R.L. Wilson (r), and the publisher, S.P. Fjestad, shown displaying the 21st Edition *Blue Book of Gun Values* cover featuring a Beretta during the 2001 SHOT Show in Las Vegas.

Preface & Acknowledgements

It hardly seems like a decade has passed since R.L. "Larry" Wilson, S.P. "Steve" Fjestad, and I first laid the groundwork for *Colt Blackpowder Reproductions & Replicas*, an unprecedented book on contemporary black powder reproductions and replicas and the predecessor to this publication. In that time, hundreds of fine percussion revolvers and cartridge conversions have passed through my hands, thousands of photographs have been taken, and I have traveled everywhere from Fredericksburg, Texas, to Gardone, Italy, in order to record the history and manufacturing of today's fine replica black powder arms.

Although we have to call this "work," dressing up in fine, period western clothing or Civil War uniforms, test firing hundreds of black powder guns, following Civil War battle reenactments, and photographing more than 150 pistols in the studio for this book has been an absolute treat. For those who enjoy collecting and shooting vintage firearms, or participating in Civil War/Old West reenactments or Single Action Shooting Society competitions, the inner child in every cowboy and Western enthusiast will find something in the pages of this book to rekindle a memory. The idea for *Black Powder Revolvers – Reproductions & Replicas* and its predecessor really began with a question I asked Larry Wilson more than a decade ago. "Where do Colt black powder pistols come from?" Now, for many shooting enthusiasts that thought probably never crossed their minds. You buy the gun, you load it and you shoot it. And it didn't matter whether the name on the pistol was Colt, Pietta, Uberti, or Euro Arms, or whether it was made in Hartford, Connecticut, Brooklyn, New York, or Brescia, Italy. On the other hand, there were those who did ask the question, and the answers in the early 1990s were at best vague and almost always wrong. One gun shop owner told me that the Second Generation Colts built in the 1970s and 1980s used the tooling from the original Colt percussion pistols. That's a pretty good tale, but it is just a tale. The original tooling had long ceased to exist, having been destroyed in the Colt factory fire of February 4, 1864, which razed the entire Hartford facility. Later tooling for the guns produced through the early 1880s was simply discarded over time. None of it survived for the production of replicas. The first Colt-style percussion pistols were created by Val Forgett, Sr., and Aldo Uberti in the 1950s, and Italy, not Hartford, Connecticut, is where the story of contemporary Colts begins.

Author Dennis Adler (c) with Colt historian and fellow author R.L. Wilson (l), and Colt Blackpowder Arms Company President Anthony Imperato, during the presentation of the A.A. White-Andrew Bourbon engraved 2nd Generation Colt Dragoon to Adler in 1997.

The very first black powder reproductions were patterned and handcrafted from original guns, as have been every replica black powder revolver produced in the past half century. As for the difference in spelling, "black powder" is the generic nomenclature and the proper spelling, "Blackpowder" is the Colt trademark name.

In the very early years of the hobby, quality was not a great issue, and considering the improved metallurgy and manufacturing techniques of contemporary gunmakers, even the least expensive replica was stronger and better built than a century-old original. With the passing of years however, and increased demand, quality has further improved to meet the more discerning requirements of customers, many of who are movie and television studio prop departments. Movies have created the greatest market for reproduction pistols. The most vivid images I can recall of the great Colt percussion revolvers were in early Clint Eastwood Westerns such as *The Good, The Bad, And the Ugly*, and *The Outlaw Josey Wales*. Eastwood probably did more to heighten awareness of these vintage Colt percussion pistols than any single individual in film making history. Eastwood's attention to period firearms was all in the name of authenticity, and that is what this book is about, the creation of the world's finest reproduction pistols. It is also about learning the value of authenticity and understanding what separates the various levels of fit, finish, and embellishment among reproductions and more importantly, those bearing the Colt name and signature.

While the onus for a book always falls upon the author, it requires the efforts and cooperation of many individuals to create the final product. With that in mind, the author would like to express his gratitude to the following individuals and companies for their contributions to the original edition of *Colt Blackpowder Reproductions & Replicas* as well as this latest volume *Black Powder Revolvers – Reproductions & Replicas*. The late Lou Imperato, Chairman of Colt Blackpowder Arms Co., and Anthony Imperato, President of Colt Blackpowder Arms Company; Giuseppe, Alessandro and Alberto Pietta of Fratelli Pietta; the late Val Forgett, Sr., founder of Navy Arms Company, Ridgefield, New Jersey; Val Forget III; Bill Ruger Jr., Sturm, Ruger & Co. Inc.; Colt Blackpowder pistol collector and historian Dennis Russell; my friend and talented gunmaker Bob Millington of ArmSport LLC in Platteville, Colorado; the brilliant master engravers John J. Adams, Sr. and John Adams, Jr.; engraver and silversmith Conrad Anderson; Mike Harvey of Cimarron F.A. Co. in Fredericksburg, Texas; Tammy Loy and Sue Hawkins of Taylor's & Co., Inc., in Winchester, Virginia; Dr. Joseph A. Murphy; master holster makers Jim Lockwood of Legends in Leather in Prescott, Arizona, and Jim Barnard of Trailrider Products in Littleton, Colorado; friends Mark McNeely and Chuck Ahearn of Allegheny Trade Co., Duncansville, Pennsylvania; and the Autry Museum of Western Heritage in Los Angeles, California, for allowing us access to the George A. Strichman collection. Without the help of these individuals, books like this would be all hat and no cattle.

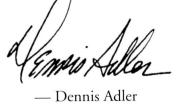

— Dennis Adler

Once a cross draw always a cross draw. The author at age 3 and at a somewhat older stage of life. When you grow up on Hopalong Cassidy, Red Ryder and The Lone Ranger, what can you expect?

An early influence was William Boyd, aka Hopalong Cassidy, seen here with the author, in appropriate black hat, at Boyd's San Fernando Valley ranch in 1951.

Author Adler (far right) discusses a photo layout with the men of the 43rd Virginia during a Civil War reenactment at Old Bedford Village, in Bedford, Pennsylvania.

CHAPTER ONE: Colt Percussion Pistols

Framed by the ornate rococo styling of an 18th century Chippendale desk are Colt Blackpowder Arms commemorative models: 1842 Paterson, 1860 Tiffany-style Army, and the massive 1851 Cochise Dragoon.

COLT PERCUSSION PISTOLS

From Paterson to Police and Navy Models

A student of repeating firearms design, which date as far back as the mid-16th century, Samuel Colt devised his own mechanism for a multiple shot revolver in 1830-31, subsequently developing a series of prototypes through 1835 with gunmaker John Pearson of Baltimore. Colt was just 21 years old when he filed his first patent for the design of a percussion revolver with mechanically-rotated cylinder. What is interesting is that he filed this patent in Great Britain, and then in France, before applying for a patent in the United States. Colt knew that a U.S. patent would preclude the filing for patents in England and France, whereas no such stipulation prevented him from filing in the U.S. after he had secured his foreign patent rights. Colt's British patent was issued on October 22, 1835, the French patent on November 16, 1836, and his U.S. patent on February 25, 1836, (followed by a letter of extension) thus providing him with the exclusive rights to build percussion revolvers based upon the fundamentals of his design through 1857.

The new pistols were known as Paterson models, named after the city in which they were produced, Paterson, New Jersey, where Samuel Colt had established the Patent Arms Manufacturing Company in 1836. The first model, a .28 caliber pocket pistol, also referred to as a Baby Paterson, was introduced in 1837 and inspired in Colt an inclination toward small, easily carried handguns which he would return to time and again throughout his career.

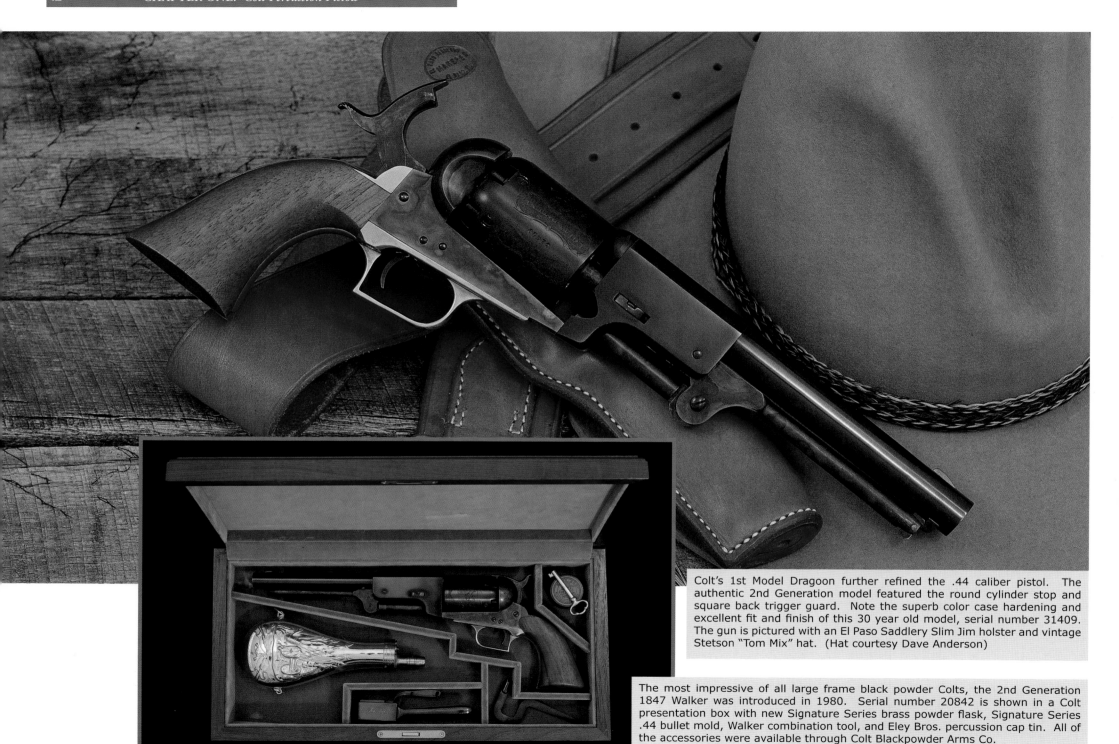

Colt's 1st Model Dragoon further refined the .44 caliber pistol. The authentic 2nd Generation model featured the round cylinder stop and square back trigger guard. Note the superb color case hardening and excellent fit and finish of this 30 year old model, serial number 31409. The gun is pictured with an El Paso Saddlery Slim Jim holster and vintage Stetson "Tom Mix" hat. (Hat courtesy Dave Anderson)

The most impressive of all large frame black powder Colts, the 2nd Generation 1847 Walker was introduced in 1980. Serial number 20842 is shown in a Colt presentation box with new Signature Series brass powder flask, Signature Series .44 bullet mold, Walker combination tool, and Eley Bros. percussion cap tin. All of the accessories were available through Colt Blackpowder Arms Co.

The larger Patersons in .34 and .36 caliber, the Belt Model No. 3 and Holster Model No. 5 (also known as the Texas Paterson) were the guns that launched Colt's business, although only briefly, as his first venture failed in 1842. This could be akin to Henry Ford's first automobile company, which failed in 1903. Ford would return to establish himself as the foremost manufacturer in his field with the Model T. Colt would make his comeback in the firearms industry in 1847. Interestingly, one of Colt's later employees was an engineer named Henry Martyn Leland, who would go on to found Cadillac shortly after the turn of the century. It was Leland's years of experience with Colt that led to his penchant for precision manufacturing, a hallmark of early Cadillac motor cars.

For Sam Colt, the Holster Model No. 5 had sown the seeds for his eventual revival in 1847, and the unparalleled success that followed. The last model developed by the Patent Arms Manufacturing Company under Colt's ownership, the No. 5, was purchased in quantity by the Texas Rangers and U.S. Mounted Riflemen in 1839 and 1840. While the Patent Arms Mfg. Company had proven unsuccessful, the No. 5 Holster Model, particularly those with attached loading levers produced in 1840 and 1841, kept interest alive in the Colt design throughout the interim years leading to development of the .44 caliber Walker Colt in 1847. In the history of Colt authorized reproductions, it is interesting to note that the Paterson model was never reproduced until 1998, making the first Colt model the very last to be commemorated.

The model which is most revered by collectors of both original Colts and Colt reproductions is the 1847 Walker. Up until the introduction of the Smith & Wesson .44 Magnum, the Walker was the most powerful handgun in the world. Like the original Walker, the 2nd and later 3rd Generation Walker models fired a .44 (actually .454) caliber ball charged with up to 60 grains of black powder delivering a round at a velocity of 1,200 feet per second backed by a force of 450 foot pounds. It was an enormous contrast to the Paterson No. 1 and No. 2 models, pistols so small they could fit in your trouser pocket. The Walker was so large that most were carried on horseback in military pommel holsters or civilian pommel-bag holsters. Those who chose to carry a Walker on their person either wore a heavy California pattern belt holster or a wide sash. The means of carry notwithstanding, there was nothing quite as impressive looking as a Walker Colt, and nothing has changed since 1847.

HOW THE FIRST COLT WALKER REPRODUCTIONS WERE PRODUCED
By R.L. Wilson

Tom Thornber headed up Colt's marketing of the new Blackpowder program. He was not an expert on antique arms, but was brilliant at promotion of any good product, and was a keen and active shooter. We were very good friends, and I bluntly told Tom that the Walkers thus far made by the Italian and Belgian gunmakers (and all other mass-produced Walkers made to date), did not hold a candle to the originals.

One day Tom called to say that Angelo Buffoli, of Armi San Marco, and one of his engineers, were coming to the Colt factory. Did I have a Walker Colt that would be available for their examination? At the time I had the finest known military Walker, E Co. No. 120, as well as C Co. No. 40, a good example with a Mormon history, which turned up in Utah. I offered to bring both revolvers to the Colt factory, at the West Hartford office (where the engineers were located). This allowed Angelo and his engineer a detailed examination; not only did they study these revolvers thoroughly, but C Co. No. 40 was taken completely apart. Each and every part was then put on a spectrograph - a means of making supremely detailed mechanical drawings and measurements. Because of the value of E Co. No. 120, Angelo and his colleague were not able to take that completely apart, but I did pull the barrel, and the cylinder, which permitted a close up look at several important details.

As was Colt's practice, some details were purposely made at a variance with the original revolvers, but the result was the first really fine quality of Walker Colts, an accurate continuation of the originals, by the company descended directly from the original manufacturer. Everyone was very proud of the results. Colt still is, and rightfully so.

CHAPTER ONE: Colt Percussion Pistols

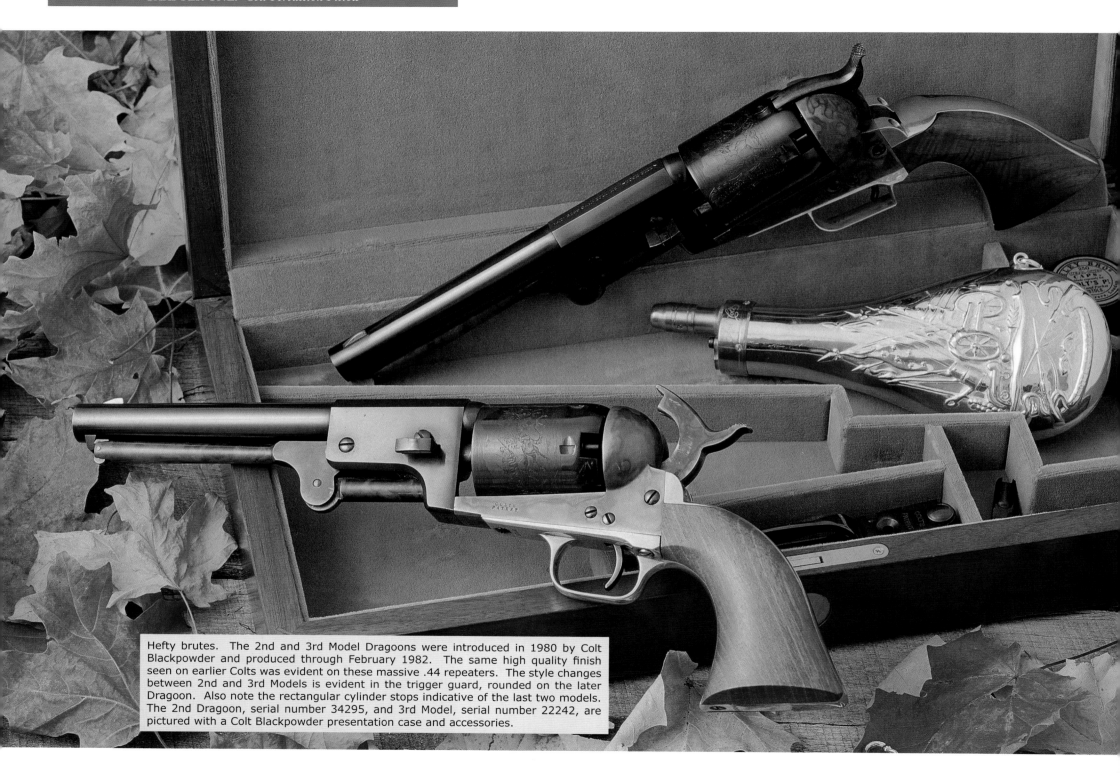

Hefty brutes. The 2nd and 3rd Model Dragoons were introduced in 1980 by Colt Blackpowder and produced through February 1982. The same high quality finish seen on earlier Colts was evident on these massive .44 repeaters. The style changes between 2nd and 3rd Models is evident in the trigger guard, rounded on the later Dragoon. Also note the rectangular cylinder stops indicative of the last two models. The 2nd Dragoon, serial number 34295, and 3rd Model, serial number 22242, are pictured with a Colt Blackpowder presentation case and accessories.

As a Colt reproduction, the 1847 Walker was first issued in June 1980 as the Colt Heritage, a limited, cased model complete with a special leather-bound and gilt edged edition of R.L. Wilson's book *The Colt Heritage*. A total of 1,853 were produced through June 1981. Each Heritage Walker was serialized from 1 to 1853 with the *Colt Heritage* book, signed by Wilson and numbered to match the gun. A striking presentation, the Walker was set into a French fit case with the book placed in a removable shelf, allowing both the pistol and Colt Heritage book to be displayed.

The large frame .44 Model Walker Heritage with hinged loading lever, had a 9 1/2 inch half round, half octagonal barrel, steel backstrap, brass square back trigger guard, US 1847 stamped above the wedge screw, and gold engraved portraits of Capt. Samuel Walker and Col. Samuel Colt on the left side of the barrel lug, with the scrollwork Heritage Banner engraved on the right. Standard Walker models were added to the Colt Blackpowder line in June 1980 and produced through April 1982. The total number of 2nd Generation guns, exclusive of the Heritage, was 2,818, with serial numbers 1200 to 4120, and 32256 to 32500, the later series produced from May 1981 through September 1981.

The Colt Blackpowder brochure quoted the provenance of the Walker as follows:

In 1846 Captain Samuel H. Walker was on the East Coast recruiting soldiers for service in the Mexican War. He had been a Texas Ranger and subsequently joined the U.S. Mounted Riflemen, also known as the Dragoons. Walker sought out Colt in New York, and stated that if he could make an improved version of the Paterson revolver, a government order would ensue for a substantial quantity. Colt and Walker collaborated on designing a 6 shot, .44 caliber, 4 lb. 9 oz. single action handgun with a 9" barrel and 2-7/16" length cylinder. To honor the support and cooperation of his co-designer, Colt dubbed the revolutionary new arm the "Walker Revolver."

An enthusiastic U.S. government contracted for 1,000 specimens to be marked individually for the Dragoon companies, A, B, C, D, and E. One hundred extra, in the serial number range 1000 to 1100, were made at Colt's own expense, for civilian sales and presentations. All 1,100 guns were completed in 1847, and bore that date; assembly was at the armory of Eli Whitney, Jr., Whitneyville, Connecticut.

The Colt Walker became a legend in its own time, and returned Sam Colt to the gunmaking business in a dramatic and decisive manner. The immediate success of the Walker enabled Colt to set up manufacturing facilities in his hometown of Hartford, in 1847. Aided by the patronage of the government and the demands of Frontier America, new models were quickly introduced.

Thus it was with the Walker that the future of Sam Colt and the cap and ball revolver was cast. For Colt Blackpowder, however, the Walker came late in its 2nd Generation series. Nearly a decade earlier, the company had introduced its own reproduction of Sam Colt's most popular revolver, the 1851 Navy.

It was no coincidence that Val Forgett, Sr. and Aldo Uberti had chosen this model in the late 1950s as the first Colt percussion revolver to be reproduced in Italy. This was the legendary six-shooter carried by James Butler Hickok, and the handgun used by more Union and Confederate soldiers during the Civil War than any other. It has also been regarded as the most beautifully designed firearm in history, an elegant combination of octagonal barrel and gracefully curved backstrap, a revolver that fell easily into hand and had superb balance.

The original 1851 Navy ranked second only to the 1849 Pocket Model in total production. It was designed by Colt to be an intermediate-sized pistol, positioned between the mammoth .44 caliber Dragoons, which followed the Walker in 1848, 1850, and 1851, and the compact 5-shot .31 caliber Pocket model. Chambered for .36 caliber, the 6-shot Navy had a barrel length of 7 1/2 inches and weighed only 2 lbs. 10 oz. It was an elegant handgun.

While the gun was frequently used by Naval officers, it was not adopted as an official sidearm for the U.S. Navy. The name actually came from the roll engraved cylinder scene depicting the Battle of Campeche, which took

CHAPTER ONE: Colt Percussion Pistols

A pair of first 2nd Generation Colt 1851 Navy models are shown with Union Officer's jacket and hat and photo of Col. Joshua Lawrence Chamberlain, the commander who turned the tide at Gettysburg in the Battle of Little Round Top. Navy models are serial numbers 9021, produced in 1971, and 14909 produced in 1977. Both feature silver plated backstraps and square back trigger guards.

Rare among the 2nd Generation Colts are the engraved Cavalry Sets with shoulder stock. These limited editions were "Class A" and "Class B" engraved. The cased set came with engraved shoulder stock and accessories. Only 23 sets were completed. This is part of No. 11. The sets are worth upwards of $6,500 today.

place May 16, 1843 between the naval forces of the Republic of Texas and Mexico. Extremely popular during the Civil War among both U.S. Army and Confederate troops, the 1851 Navy was also copied by Southern gunmakers because of its light weight and hard hitting .36 caliber round.

Colt continued to offer the 1851 Navy throughout the percussion era, producing some 215,000 guns in Hartford, and another 42,000 in England. Today, the 1851 Navy is the easiest original Colt to acquire. Examples in 98% condition, however, can command up to $25,000 and even modest 50% guns average from $2,500 to $10,000 depending upon the model variation.

One hundred and twenty years after Samuel Colt introduced the 1851 Navy, the Colt Blackpowder Arms Co. produced its first reproduction. The 2nd Generation model was offered in two series, C and F, and in three initial versions.

First was the standard 1851 model with serial numbers 4201 through 25099, produced from 1971 through 1978. A second series was introduced in May 1980 and manufactured through October 1981 with serial numbers 24900 through 28850. Finally, there was a series of limited editions beginning with the 1971 Civil War commemoratives. These were the Robert E. Lee, wood cased with accessories, the Ulysses S. Grant, wood cased with accessories, and 250 "Blue and Gray" double cased sets. A total of 4,750 Grant and Lee Navys were produced and 25,150 standard Navy models.

The Grant and Lee sets are the most desirable high production models, and each was specially cased with period-style accessories. The U.S. Grant model in .36 caliber came in a blue satin and felt lined presentation box with pewter powder flask, a bullet mold, and percussion cap tin. The commemorative revolver, with hinged and latched loading lever, 6-shot roll engraved cylinder, 7 1/2 inch octagonal barrel, silver-plated backstrap and square back trigger guard, had the engraved N.Y. Address on top of the barrel, and Ulysses S. Grant Commemorative Nineteen Seventy One engraved on the left side of the barrel.

The Lee model was cased in gray satin and felt and was further differentiated from the Grant Navy by having the later round trigger guard, a pewter Colt-style powder flask, combination tool, and Robert E. Lee Commemorative Nineteen Seventy One engraved along the left side of the barrel.

The elegantly styled 1860 Army has become the most reproduced Colt model of all time. Here, a pair of 2nd Generation models, serial numbers 209154 and 205933, are pictured with a Union Army cavalry jacket, kepi, and saber.

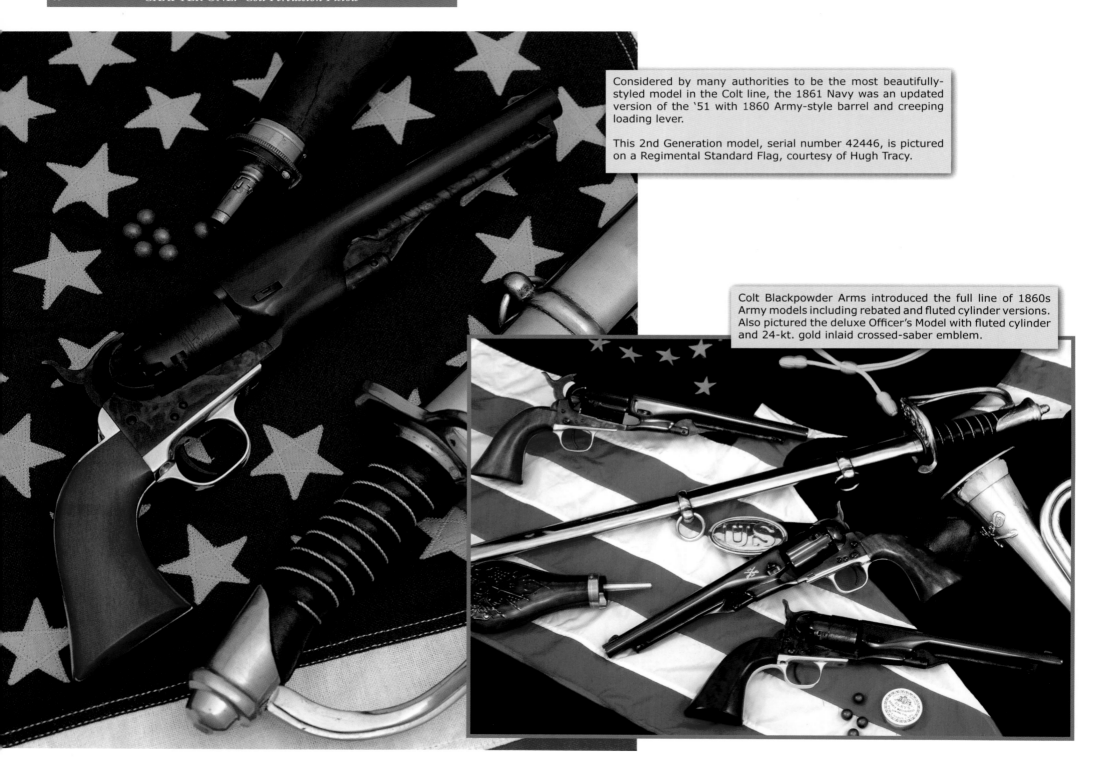

Considered by many authorities to be the most beautifully-styled model in the Colt line, the 1861 Navy was an updated version of the '51 with 1860 Army-style barrel and creeping loading lever.

This 2nd Generation model, serial number 42446, is pictured on a Regimental Standard Flag, courtesy of Hugh Tracy.

Colt Blackpowder Arms introduced the full line of 1860s Army models including rebated and fluted cylinder versions. Also pictured the deluxe Officer's Model with fluted cylinder and 24-kt. gold inlaid crossed-saber emblem.

Both the individual cased models and the limited edition double cased sets are considered collectible today with values as high as $2,500 for 100% examples of the cased pair. The double gun set, however, is a rare find, according to noted Colt Blackpowder collector Dennis Russell.

In addition to the standard production models, Colt Blackpowder manufactured 300 Navys with blank cylinders (no roll engraved battle scene) within the serial number range 28851 to 29150, and 490 examples with a stainless steel finish in the serial number range 29151s through 29640s. Model number F1110 Z was produced in stainless steel with ivory stocks, and F1110 in stainless with B engraving and ivory stocks [1]. The Colt factory also produced a very limited run of 174 engraved models known as the Conquistador Special Edition in 1976,[2] and 500 Navys with polished brass backstraps and trigger guards, through 1978. A series of 50 Navy revolvers embellished with gold bands and "B" engraving, were produced by the Colt Custom Shop in 1985 and 1986.

Among the more rare of 2nd Generation Colt Blackpowder models were the stainless steel 1851 Navy and 1860 Army. These are still a rare find today and command a premium. There are also a handful of even rarer 1861 Navy stainless models all of which are now in private collections.

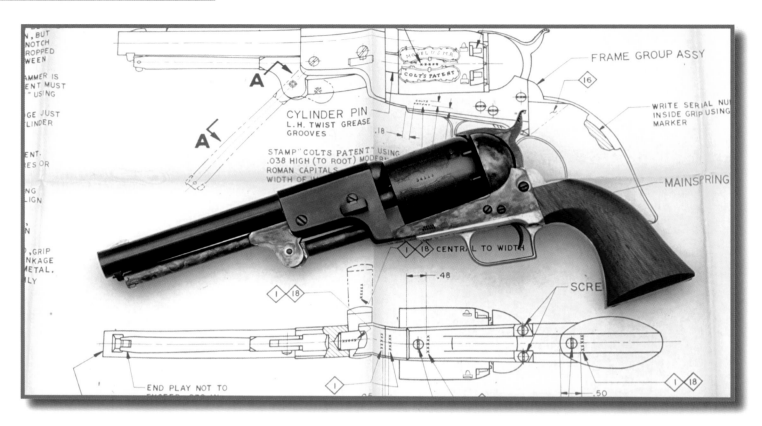

THE .44 CALIBER GUNS

In 1974 Colt added another model to the Blackpowder line, a 3rd Model Dragoon, which was produced through 1978 in the serial number range 20901 through 25099. Colt also built 25 prototype 3rd Model Dragoons with serial numbers 20801 through 20825. These would be considered very rare today.

Somewhat out of historical sequence, Colt skipped the 1st and 2nd Model Dragoons (later introduced in 1980) and next brought out the popular 1860 Army model in November 1978.

Sam Colt had waited almost a decade to improve upon the 1851 Navy but when he did, it was another high watermark in firearms design. The 1860 Army was nearly the same size as the venerable '51 Navy, but packed a whopping .44 caliber round with nearly as much punch as the heavier Dragoons. Colt used the same basic frame as the 1851 Navy, but with a slightly longer backstrap and grip, a new rebated cylinder (milled larger in diameter approximately 3/4 of an inch forward of the breech to allow for the larger caliber), and a beautifully contoured round barrel measuring 8 inches in length. Bearing the same roll engraved battle scene as the 1851 Navy, it was a handsome looking firearm and an immediate success for Colt. This has by far become the second most popular black powder reproduction in Colt history, and aside from the 1851 Navy, it is the most copied percussion revolver of all time. A second version of the 1860 Army was introduced with a fully fluted cylinder, and both models saw extensive use during the Civil War.

As noted by Colt historian R.L. Wilson, "Approximately 129,000 Model 1860 revolvers were issued to U.S. troops for Civil War service, several thousand of them equipped with an attachable shoulder stock, an accessory to allow firing the arm as a carbine. More 1860 Armys were purchased by the U.S. government than any other model of Colt or any other make of black

1860 style with rebated cylinder was manufactured from November 1978 to November 1982 in serial number sequence 201000 through 212835; also with an electroless nickel finish in 1982; in a cased limited edition of 500 guns in 1979; with a fluted cylinder from July 1980 through October 1981 in serial number range 207330 to 211262; and in stainless steel from January 1982 to April 1982 with serial number sequence 211263s to 212835s.

Colt also produced a number of special edition Army models. One series was commissioned by the Hodgdon Powder Company in 1979 to commemorate the Butterfield Overland Stage. This was limited to 500 guns and came with an extra cylinder in a French fit book-style case. A dozen 1860 Armys were specially finished in nickel plate and fitted with ivory stocks in 1984. There were 3,001 U.S. Cavalry 200th Anniversary double pistol sets cased with a shoulder stock and accessories produced in 1977 along with a limited edition of cased engraved sets with matching hand engraved accessories. In 1979 a series of 500 cased 1860 Army models were built, and in 1980 a special Interstate Edition of 200 guns.

The Colt 1862 Pocket Model of Navy caliber and 1862 Pocket Police were the next additions to the Blackpowder line, introduced in December 1979 and January 1980, respectively. The original guns were scaled down versions of the 1851 Navy and 1860 Army and the last percussion models introduced by Colt prior to the 1873 Peacemaker. Notes Wilson, "Both actually appeared in 1861, just months before the inventor's death, on January 10th 1862. Stocks and frames were identical on these revolvers, as was their serial range, caliber and number of shots (.36, 5-shot rebated cylinder), and barrel lengths, (4 1/2", 5 1/2", and 6 1/2")."

The 1862 Pocket Police was distinguished by its fluted and rebated cylinder, round 1860 Army-style barrel, and creeping lever for loading. The Pocket Model of Navy Caliber (the correct designation for this model) featured a rebated round cylinder, roll engraved with stagecoach holdup scene, and octagonal barrel with hinged-type loading lever.

"The .36 caliber chambering of these medium size revolvers made them highly prized pocket sidearms. As also true with the 1849 Pocket, a number were carried by Civil War soldiers as backup to their single shot muskets," noted Wilson in a Colt Blackpowder sales booklet published in 1978. That 30-year-old booklet is itself a collectable item in mint condition.

Both the 2nd Generation 1862 Pocket Model of Navy Caliber and Police were produced in a limited series of 500 cased models complete with accessories. The cased sets were manufactured in 1979 and 1980.

powder revolver. This was the staple handgun of the Civil War, and played the same role in the Plains Indian wars, until succeeded by the Colt Peacemaker .45, brought out in 1873." Some 200,500 1860 Armys were manufactured, making it the second highest production Colt up to that time.

The Colt Blackpowder 2nd Generation reprise of the 1860 Army remained in production until 1982, and was offered in a variety of models. The original

A small gun with a hardy muzzle blast, the 1862 Pocket Model of Navy Caliber was a scaled down 5-shot version of the venerable .36 caliber 1851 Navy. The 2nd Generation models were produced from December 1979 through November 1981. The gun pictured, serial number 50611, was among 500 limited edition cased revolvers. Shown with Colt pewter powder flask c.1971 and a hand tooled flap holster.

The 2nd Generation Pocket Model of Navy Caliber was produced through November 1981 and the Pocket Police through September 1981 with serial numbers 48000 to 58850 for the Navy and 49000 to 57300 for the Police. Both Pockets were also produced in a limited edition of 500, each in a French fitted presentation box with bullet mold, pewter powder flask, percussion cap tin and combination tool. The beautifully-styled presentation boxes were affixed with a brass medallion mounted in the lid featuring a cast bust of Samuel Colt and the legend COLT AUTHENTIC BLACKPOWDER. The series was produced in 1979 and 1980 within the production serial number run. In addition, a very limited edition of 25 Navy and Pocket Police revolvers were produced in nickel finish with ivory stocks in 1984. These would become the rarer of 2nd Generation Pocket models.

The 1980 Colt Blackpowder series featured yet another historic reproduction, the 1861 Navy, which was added in September and produced through October 1981 in the serial number sequence 40000 to 43165. The original 1861 Navy was Colt's update on the 1851 version, and was again chambered in .36 caliber. The new model was virtually identical to the 1860 Army, save for the rebated cylinder, which was unnecessary for the smaller caliber chambering. Interestingly, Colt did not discontinue the 1851 Navy when the 1862 was introduced, and both remained in production through 1873. Only 38,000 Model 1861 Navys were manufactured.

The Colt Blackpowder 2nd Generation 1861 Navy followed the original Colt formula, fitting the 1851 frame and cylinder with a streamlined round barrel and replacing the hinged loading lever with the later creeping style used on the 1860 Army. For sheer style and balanced proportions, the 1861 Navy is often considered to be the most beautifully designed Colt percussion revolver.

Within the 2nd Generation 1861 Navy series is one of the rarest models produced by Colt. In 1982 a total of just six stainless steel 1861 Navy

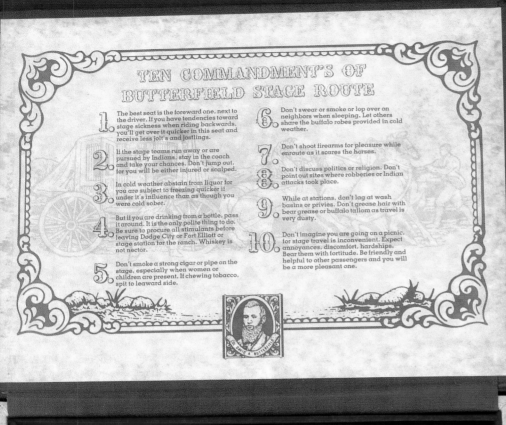

Commissioned by the Hodgdon Powder Co. the 2nd Generation Butterfield Overland Despatch model 1860 Army came with a cut down 5 1/2 inch barrel and engraved cylinder picturing the Butterfield stagecoach scene and a spare cylinder engraved with a portrait of Col. David Butterfield, a Conestoga wagon and longhorn steer. The barrel was engraved Butterfield Overland Despatch. The edition was limited to 500 and came in a French fit book-style case. The example pictured with gold embellished scenes (the guns have been seen either in blue or with gold filled engraving) is serial number 201323.

models were built with serial numbers 43166s to 43171s. This would become the rarest of all 2nd Generation Colts.

A new line of 1st, 2nd, and 3rd Model Dragoons were added in 1980, all of which were introduced in January. The three versions were produced in the same serial number series beginning with 25100 and concluding in February 1982 with serial number 34500.

Historically, the 1st Model Dragoon was a slightly scaled down and improved version of the 1847 Walker, which had proven to be cumbersome in the field and troubled by a failing loading lever latch. This was merely a post with a widened tip protruding from the bottom of the barrel through the base of the loading lever, and designed to catch on the underside of the hinge. This worked until the tip began to wear, after which the lever could be knocked free by recoil, allowing it to drop and the plunger to impede cylinder rotation. On the Dragoon, the barrel was shortened from 9 inches to 7 1/2 inches and the cylinder length reduced to 1 7/16 inches. The overall weight was pared down by nearly half a pound, and a new spring-tensioned catch at the front of the ramrod eliminated the loading lever problem. There were also some minor improvements to the internal mechanism and the grip was made more substantial for the heft of the gun. There was also an earlier "interim" version produced in 1848 and known as the Whitneyville-Hartford Dragoon, which resembled a cross between a Walker and a 1st Model Dragoon. These were not reproduced until the 3rd Generation Colt series in the 1990s.

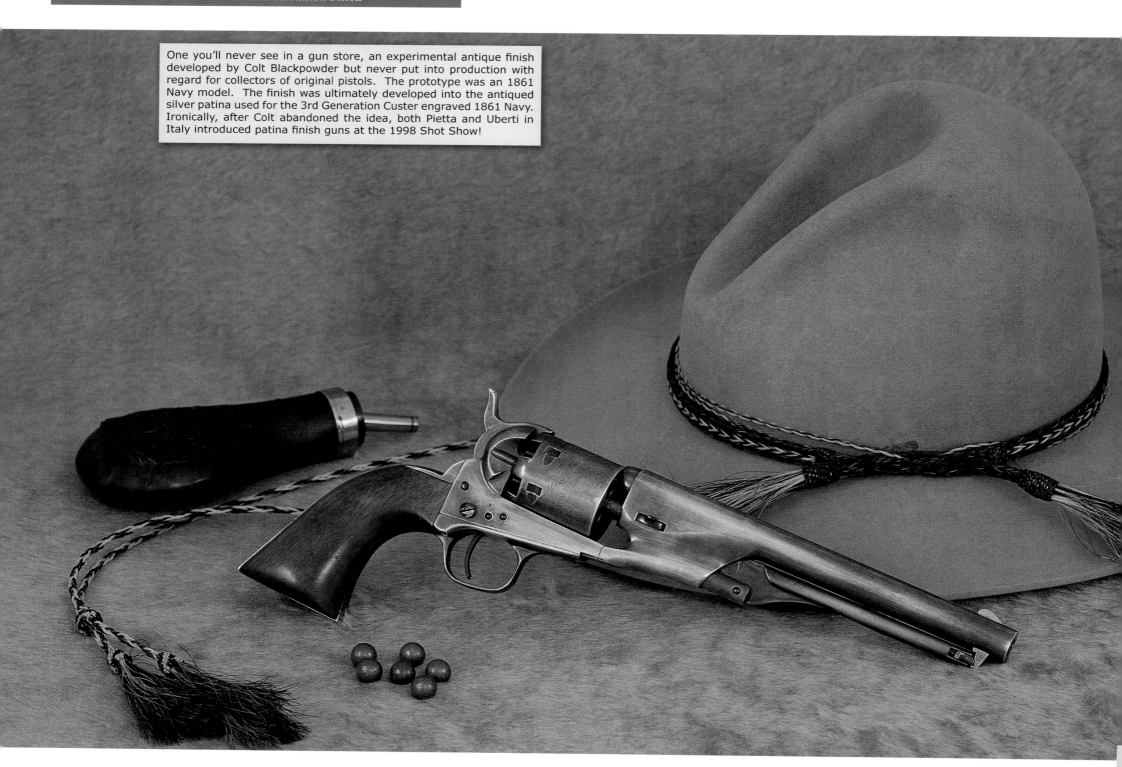

One you'll never see in a gun store, an experimental antique finish developed by Colt Blackpowder but never put into production with regard for collectors of original pistols. The prototype was an 1861 Navy model. The finish was ultimately developed into the antiqued silver patina used for the 3rd Generation Custer engraved 1861 Navy. Ironically, after Colt abandoned the idea, both Pietta and Uberti in Italy introduced patina finish guns at the 1998 Shot Show!

Eight limited edition Dragoons were also produced, the first in 1976, a 3rd Dragoon Bicentennial Commemorative with serial numbers 0001DG through, appropriately, 1776DG. In 1978 a series of 52 Statehood 3rd Models were built representing various states. In 1981, a group of 100 1st Model "Yorktown" Dragoons in serial number sequence 32500 to 32999 were produced. A series of 200 "Garibaldi" 3rd Model engraved Dragoons in serial number sequence GCA1000 to GCA1199, were manufactured in 1982. In 1984 a run of 75 High Polish blue 3rd Models were built, an additional five with ivory stocks, and 20 with nickel finish and ivory stocks. A series of 50 B engraved 3rd Models were produced in 1985-86. Additionally, in 1988 there were 25 Sampler 3rd Models done in blue and 25 finished in nickel. The 3rd Model Dragoon was by far the most varied of all 2nd Generation Colts.

The last model to be introduced by Colt Blackpowder was the 1847-48 Baby Dragoon, a pocket-sized version of the 1st Model Dragoon, and a throwback to the very first Paterson No. 1, chambered for .31 caliber, limited to 5-shots, and conspicuously absent of a loading lever. (In its place, the cylinder pin doubled as a loading ramrod). Standard barrel lengths for the original guns ranged from 3-inches to 6-inches. The 4-inch model was the most popular and that was the barrel length chosen for the 2nd Generation Colt, introduced in 1979 as a limited edition of 500 presentation models each in a French fitted case with powder flask, bullet mold, percussion cap tin and combination tool. Between February and April 1982 another 1,352 Baby Dragoons were produced for general sales. The entire serial number range from 1979 through 1982 was 16000 to 17851.

Another rare variation, a Custer Commemorative in polished nickel. This was a variation that was not put into production, and like the antique finish is a one-of-a-kind example.

The distinguishing characteristics between the three Dragoon models are minor, however, one can easily spot a 1st Model by its square back trigger guard and oval cylinder stop slots. The 2nd Model has rectangular stop slots, and 3rd Model Dragoons have rectangular stop slots and a rounded trigger guard. All three versions share a half round, half octagonal barrel, 2-3/16 inch cylinder and roll engraved Texas Ranger and Indian fight scene.

The question that arises whenever collectors gather is whether or not Colt actually built the 2nd Generation Blackpowder models. The answer is yes, and no.

When Colt Industries decided to revive the 1851 Navy they went to Val Forgett and Navy Arms. Forgett saw to the production of the cast and machined parts in Italy, which were shipped to the United States and completed by Colt in the Hartford factory. All of the early 1851 Navy models produced in 1971 and 1972, including the commemorative U.S. Grant and Robert E. Lee sets, were produced through Forgett, as well as the early Dragoons.

As Colt continued to add new models, the Blackpowder line was produced in Middlesex, New Jersey, on a sub contract basis by Iver Johnson Arms and Louis Imperato. The same process of finishing components produced in Italy to Colt's specifications continued throughout the entire 2nd Generation, however, the frames, center pins, nipples and screws for all 2nd Generation models were made by Iver Johnson Arms in New Jersey.

So is a 2nd Generation Colt a Colt? Is a Ford completed in Detroit from parts made in Canada and Mexico a Ford? The answer is yes, and a Colt pistol bearing the Colt name and patent, regardless of where the individual components are cast, is a Colt.

The 2nd Generation Blackpowder pistols, while numbering in the thousands, represent but a fraction of original Colt production. Each is an authentic Colt revolver with all of the history and provenance of the original 19th century models, contemporary collectables that are today assuming their rightful place alongside the originals, adding yet another chapter to the Colt Heritage.

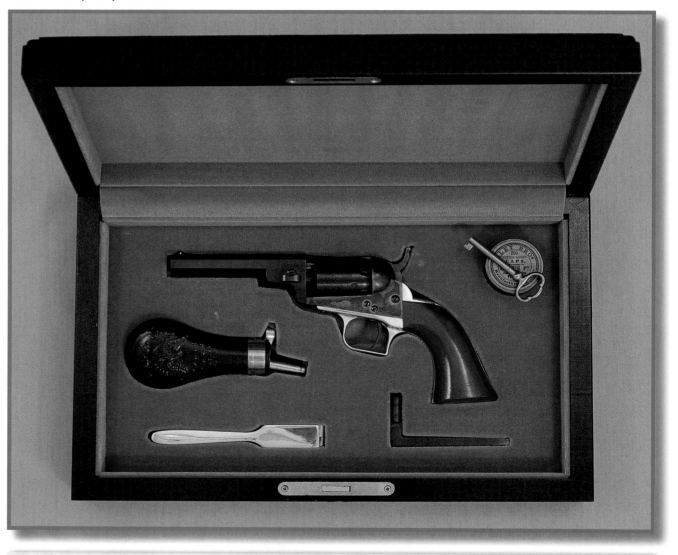

A total of 1,852 Baby Dragoons were made which included a series of 500 presentation models in a French fit case with accessories. The model was reprised by Colt Blackpowder in 1997, once again with a 4 inch barrel, silver backstrap and trigger guard.

Colt Blackpowder 2nd Generation Model Specifications

.44 **Model 1847 Walker**, Revolver, with Hinged Loading Lever, 6 Shot, 9 1/2 inch Half Round, Half Octagonal Barrel, Steel Backstrap, Brass Square Back Trigger Guard, US 1847 stamped above wedge screw, N.Y. Address. Produced 1980-1982.

.44 **Model 1847 Walker Heritage**, Revolver, with Hinged Loading Lever, 6 Shot, 9 1/2 inch Half Round, Half Octagonal Barrel, Steel Backstrap, Brass Square Back Trigger Guard, US 1847 stamped above wedge screw, Gold Engraved Portraits of Capt. Walker and Samuel Colt on Left Side, Heritage Banner on Right, N.Y. Address. In Presentation Case With *The Colt Heritage* book by R.L. Wilson. Produced 1980-1981.

.44 **1st Model Dragoon**, Revolver, with Hinged and Latched Loading Lever, 6 Shot, 7 1/2 inch Half Round, Half Octagonal Barrel, Oval Cylinder Stop Slots, Brass Backstrap and Square Back Trigger Guard, N.Y. Address. Produced 1980-1982.

.44 **2nd Model Dragoon**, Revolver, with Hinged and Latched Loading Lever, 6 Shot, 7 1/2 inch Half Round, Half Octagonal Barrel, Rectangular Cylinder Stop Slots, Brass Backstrap and Square Back Trigger Guard, N.Y. Address. Produced 1980-1982.

.44 **3rd Model Dragoon**, Revolver, with Hinged and Latched Loading Lever, 6 Shot, 7 1/2 inch Half Round, Half Octagonal Barrel, Rectangular Cylinder Stop Slots, Brass Backstrap and Round Trigger Guard, N.Y. Address. Produced 1980-1982

.31 **Model 1848 Baby Dragoon**, Revolver, Pocket Pistol, no Loading Lever, 5 Shot, 4 inch Octagonal Barrel, Silver-Plated Backstrap and Square Back Trigger Guard, Two line New York address. Produced 1981.

.36 **Model 1851 Navy**, Revolver, with Hinged and Latched Loading Lever, 6 Shot, 7 1/2 inch Octagonal Barrel, Silver-Plated Backstrap and Square Back Trigger Guard, N.Y. Address. [Also Made with Brass Backstrap and Square Back Trigger Guard, or with Stainless Steel Finish] Produced 1971-1978.

.36 **Model 1851 Navy U.S. Grant Commemorative**, Revolver, with Hinged and Latched Loading Lever, 6 Shot, 7 1/2 inch Octagonal Barrel, Silver-Plated Backstrap and Square Back Trigger Guard, N.Y. Address, Barrel Engraved, Ulysses S. Grant Commemorative Nineteen Seventy One. In Blue Satin and Felt Lined Presentation Case with Pewter Powder Flask and tools. Produced 1971.

.36 **Model 1851 Navy Robert E. Lee Commemorative**, Revolver, with Hinged and Latched Loading Lever, 6 Shot, 7 1/2 inch Octagonal Barrel, Silver-Plated Backstrap and Round Trigger Guard, N.Y. Address, Barrel Engraved, Robert E. Lee Commemorative Nineteen Seventy One. In Gray Satin and Felt Lined Presentation Case with Pewter Powder Flask and tools. Produced 1971.

.44 **Model 1860 Army**, Revolver, with Creeping-Style Loading Lever, 6 Shot Rebated Cylinder, 8 inch Round Barrel, Steel Backstrap and Brass Trigger Guard, N.Y. U.S. America Address. [Also produced with Stainless Steel Finish] Produced 1978-1982

.44 **Model 1860 Army**, Revolver, with Creeping-Style Loading Lever, 6 Shot Fluted Cylinder, 8 inch Round Barrel, Steel Backstrap and Brass Trigger Guard, N.Y. U.S. America Address. Produced 1980-1981.

.36 **Model 1861 Navy**, Revolver, with Creeping-Style Loading Lever, 6 Shot, 7 1/2 inch Round Barrel, Silver-Plated Backstrap and Trigger Guard, N.Y. U.S. America Address. Produced 1980-1981.

.36 **Model 1862 Pocket Model of Navy Caliber**, Revolver, with Hinged and Latched Loading Lever, 5 Shot Rebated Cylinder, 5 1/2 inch Octagonal Barrel, Silver-Plated Backstrap and Square Back Trigger Guard, N.Y. U.S. America Address. Produced 1979-1981 and 1984.

.36 **Model 1862 Pocket Police**, Revolver, with Creeping-Style Loading Lever, 5 Shot Semi-Fluted Cylinder, 5 1/2 inch Round Barrel, Silver-Plated Backstrap and Round Trigger Guard, N.Y. U.S. America Address. Produced 1979-1981 and 1984.

The big three, Colt Blackpowder's 3rd Generation Signature Series Dragoons. Pictured from left to right, 3rd Model Dragoon civilian, 3rd Model Dragoon steel backstrap, and 3rd Model Dragoon with fluted cylinder.

THE NEXT GENERATION

Much of the new found interest in Colt percussion pistols stems from movies such as Kevin Costner's Oscar winning film *Dances With Wolves*, the two Gettysburg films (*Gettysburg* and *Gods and Generals*) and Ken Burns' excellent PBS documentaries on the Civil War and the American West.

With the values of original Colts continuing to rise throughout the last quarter century, and a significant increase in the demand for reproduction pistols used for Civil War reenactments and Cowboy Action Shooting, and a generation of new collectors, Colt Blackpowder Arms Company in Brooklyn, New York, was organized in 1993 by the manufacturers of the 2nd Generation Colt pistols, Louis and Anthony Imperato. A year later, the 3rd Generation Colt Blackpowder Signature Series was introduced.

The first model offered was the traditional 1851 Navy, followed by the 1847 Walker, 1860 Army, and 1849 Pocket. With these first guns a new tradition was established. More than a reissue of the models offered from 1971 to 1984, each was embellished along the backstrap with the flamboyant signature of Sam Colt. On steel frame guns the signature was inlaid in gold, and engraved on silver plated and polished brass backstraps. The Colt signature added a new sense of heritage to the guns, as well as a distinguishing characteristic from the 2nd Generation Colts.

By 1996, Lou and Anthony Imperato had recreated all of the original models, as well as adding two new historic Colts to the line that were never offered in the 2nd Generation, the 1849 Pocket Dragoon, literally a scaled down 1st Model Dragoon chambered in .31 caliber, and the 1862 Trapper, a pocket-sized .36 caliber revolver of which only 50 originals were ever made. A variation of the 1862 Pocket Police, the Trapper was fitted with a short 3 1/2 inch round barrel without loading lever and accompanied by a brass ramrod measuring 4 5/8 inches and designed to fit through the loading lever opening in place of the plunger.

The 3rd Generation Colts have become landmark models, expanding the variety of guns available to the collector and shooter. Within three years of their introduction, there were more than a dozen Signature Series Colts on the market, covering every caliber, size, and model produced in Hartford, including the transitional Whitneyville-Hartford Dragoon, originally produced in 1848. This was the first model assembled in Colt's new Hartford, Connecticut, plant, and only 240 were made before the 1st Model Dragoon took its place.

Introduced in 1996, the Colt Blackpowder Whitneyville-Hartford was the first historic commemorative model in the 3rd Generation. The following year Colt issued the 150th Anniversary Walker Dragoon, with gold "*A Company No. 1*" markings on the barrel lug, frame and cylinder to commemorate the Walker's use by the U.S. Mounted Rifles in the war with Mexico. The original A Company models ended with serial number 220 and the 150th Anniversary Walkers began with number 221.

Colt Blackpowder 3rd Generation Model Specifications

.36 Model 1842 Paterson, Revolver, No. 5 Holster Model with Hinged Loading Lever, 5 Shot, 7 1/2 inch Octagonal Barrel, Steel Backstrap, Period-Style Scrollwork with Punch-Dot Background on both sides of Barrel Lug, Frame and Standing Breech, Loading Lever Rod Tip, Hammer and Backstrap. Two 18-kt. Gold Bands inlaid at the Muzzle, encircling the Cylinder and one Gold Band around the circumference of the front of the Standing Breech. Royal Blue mirror finish, premium Walnut Stocks, authentic "- Patent Arms Mg.Co. Paterson, NJ - Colt's PL -" barrel address. Introduced in 1998. Standard model without engraving introduced in 2002. Produced one year only.

.44 Model 1847 Walker, Revolver, with Hinged Loading Lever, 6 Shot, 9 1/2 inch Half Round, Half Octagonal Barrel, Steel Backstrap, Brass Square Back Trigger Guard, "US 1847" stamped above wedge screw, N.Y. Address. Introduced in 1994.

.44 Model 1847 Walker, 150th Anniversary Edition, Revolver, with Hinged Loading Lever, 6 Shot, 9 1/2 inch Half Round, Half Octagonal Barrel, Steel Backstrap, Brass Square Back Trigger Guard, US 1847 stamped above wedge screw, "A COMPANY No.1" engraved on left side of Barrel, Frame, and on Cylinder, N.Y. Address. Series begins with Serial Number 221. Introduced in 1997.

.44 Model 1848 Whitneyville-Hartford Dragoon, Revolver, with Hinged and Latched Loading Lever, 6 Shot, 7 1/2 inch Half Round, Half Octagonal Barrel, Oval Cylinder Stop Slots, Silver-Plated Backstrap and Square Back Trigger Guard, N.Y. Address, "Sam Colt" Signature engraved on Backstrap. Introduced in 1996.

.44 Model 1848 Whitneyville-Hartford Commemorative "Marine" Dragoon, Revolver, with Hinged and Latched Loading Lever, 6 Shot, 7 1/2 inch Half Round, Half Octagonal Barrel, Oval Cylinder Stop Slots, Silver-Plated with 24-kt. Gold-Plated Cylinder, Backstrap and Square Back Trigger Guard, N.Y. Address, "Semper Fidelis" engraved on Backstrap. Introduced in 1997.

.31 Model 1848 Baby Dragoon, Revolver, 5 Shot, 4 inch Octagonal Barrel, Silver Plated Backstrap and Square Back Trigger Guard, N.Y. Address, "Sam Colt" Signature engraved on Backstrap. Introduced in 1998.

.44 Model 1848 1st Model Dragoon, Revolver, with Hinged and Latched Loading Lever, 6 Shot, 7 1/2 inch Half Round, Half Octagonal Barrel, Oval Cylinder Stop Slots, Brass Backstrap and Square Back Trigger Guard, N.Y. Address, "Sam Colt "Signature engraved on Backstrap. Introduced in 1998.

.44 Model 1850 2nd Model Dragoon, Revolver, with Hinged and Latched Loading Lever, 6 Shot, 7 1/2 inch Half Round, Half Octagonal Barrel, rectangular Cylinder Stop Slots, Brass Backstrap and Square Back Trigger Guard, N.Y. Address, "Sam Colt" Signature engraved on Backstrap. Introduced in 1998.

.44 3rd Model Dragoon, Revolver, with Hinged and Latched Loading Lever, 6 Shot, 7 1/2 inch Half Round, Half Octagonal Barrel, Rectangular Cylinder Stop Slots, Brass Backstrap Civilian Model with Round Trigger Guard, N.Y. Address, "Sam Colt" Signature engraved on Backstrap. Introduced in 1996.

.44 3rd Model Dragoon, Revolver, with Hinged and Latched Loading Lever, 6 Shot, 7 1/2 inch Half Round, Half Octagonal Barrel, Rectangular Cylinder Stop Slots, Steel Backstrap Military Model with Brass Round Trigger Guard, N.Y. Address, "Sam Colt" Signature engraved in Gold on Backstrap. Introduced in 1996.

.44 3rd Model Dragoon, Revolver, with Hinged and Latched Loading Lever, 6 Shot, 7 1/2 inch Half Round, Half Octagonal Barrel, Fluted Cylinder, Rectangular Cylinder Stop Slots, Silver-Plated Backstrap and Round Trigger Guard, N.Y. Address, "Sam Colt" Signature engraved on Backstrap. Introduced in 1996.

.44 Model 1851 3rd Model Dragoon, Cochise Edition, Revolver, with Hinged and Latched Loading Lever, 6 Shot, 7 1/2 inch Half Round, Half Octagonal Barrel, Gold Engraved Cylinder, Gold Backstrap and Trigger Guard, N.Y. Address, "Sam Colt" Signature engraved on Backstrap. Gold

Plated Pony Scene on Left Side of Barrel, Cochise legend on Octagonal Portion of Barrel along with Peace Pipe and Tomahawk, Gold-Plated Buffalo on Lower Corner of Frame, Blackhorn stocks with Cochise Portrait on Left Side. Introduced in 1997.

.31 **Model 1849 Pocket**, Revolver, with Hinged and Latched Loading Lever, 5 Shot, 4 inch Octagonal Barrel, Silver-Plated Backstrap and Trigger Guard, N.Y. Address, "Sam Colt" Signature engraved on Backstrap. Introduced in 1994.

.36 **Model 1851 Navy**, Revolver, with Hinged and Latched Loading Lever, 6 Shot, 7 1/2 inch Octagonal Barrel, Silver-Plated Backstrap and Square Back Trigger Guard, N.Y. Address, "Sam Colt" Signature engraved on Backstrap. Also available in London Navy Model. Introduced in 1994, London Navy in 1997. Limited edition 150th Anniversary 1851 Navy with Round Trigger Guard, hand engraved in nautical theme, with 70 percent coverage, carved ivory grips with fouled anchor, royal blue finish with 24-kt. trim. Produced from 2000-2002.

.44 **Model 1860 Army**, Revolver, with Creeping-Style Loading Lever, 6 Shot Rebated Cylinder, 8 inch Round Barrel, Steel Backstrap and Brass Trigger Guard, N.Y. U.S. America Address, "Sam Colt" Signature engraved in Gold on Backstrap. [Also issued in nickel finish, and Officer's Model with Fluted Cylinder and Embellished with Crossed Sabers and the legend US 1860, Engraved in Gold on Right Side of Barrel.] Introduced in 1994, Officer's Model in 1995, Nickel in 1997.

.44 **Model 1860 Army Gold U.S. Cavalry**, Revolver, with Creeping-Style Loading Lever, 6 Shot Gold-Plated Rebated Cylinder Engraved on Both Sides with Crossed Sabers and the legend US 1860, 8 inch Round Barrel with Double Gold Bands around the Muzzle, Gold Backstrap and Trigger Guard, N.Y. U.S. America Address, "Sam Colt" Signature engraved on Backstrap. Introduced in 1996.

.44 **Model 1860 Army**, Revolver, with Creeping-Style Loading Lever, 6 Shot Fluted Cylinder, 8 inch Round Barrel, Steel Backstrap and Brass Trigger Guard, N.Y. U.S. America Address, "Sam Colt" Signature engraved in Gold on Backstrap. Introduced in 1995.

.44 **Model 1860 Army Tiffany Revolver**, with Creeping-Style Loading Lever, 6 Shot Rebated Cylinder, 8 inch Round Barrel, Fully Covered with L.D. Nimschke Style Scrollwork, Silver and 24-kt. Gold-Plated, N.Y. U.S. America Address, Deluxe Tiffany-Style Grips plated in Pure Silver. Introduced in 1998.

.36 **Model 1861 Navy**, General Custer Edition, Revolver, with Creeping-Style Loading Lever, 6 Shot Non-Rebated Cylinder, 7 1/2 inch Round Barrel, Steel Backstrap and Trigger Guard, N.Y. Address, "Sam Colt" Signature Engraved on Backstrap. Antiqued Silver Finish, Nimschke-Style Engraving on Barrel, Loading Lever, Cylinder, Frame, Hammer, Trigger Guard, and Backstrap. Walnut Stocks with Eagle and Shield on Left, Checkered on Right. Introduced in 1996.

.36 **Model 1861 Navy**, Revolver, with Creeping-Style Loading Lever, 6 Shot Non-Rebated Cylinder, 7 1/2 inch Round Barrel, Steel Backstrap and Trigger Guard, N.Y. Address, "Sam Colt" Signature Engraved in Gold on Backstrap. Introduced in 1995.

.36 **Model 1862 Pocket Model of Navy Caliber**, Revolver, with Hinged Loading Lever, 5 Shot Rebated Cylinder, 5 1/2 inch Octagonal Barrel, Silver-Plated Backstrap and Square Back Trigger Guard, N.Y. Address, "Sam Colt" Signature Engraved on Backstrap. Introduced in 1996.

.36 **Model 1862 Pocket Police**, Revolver, with Creeping-Style Loading Lever, 5 Shot Semi-Fluted Cylinder, 5 1/2 inch Round Barrel, Silver-Plated Backstrap and Square Back Trigger Guard, N.Y. Address, "Sam Colt" Signature Engraved on Backstrap. Introduced in 1997.

.36 **Model 1862 Trapper**, Revolver, No Loading Lever, 5-Shot Semi-Fluted Cylinder, 3 1/2 inch Round Barrel, Silver-Plated Backstrap and Trigger Guard, N.Y. Address, "Sam Colt" Signature Engraved on Backstrap. Comes with Brass Ramrod. Introduced in 1995.

*****

[1] Produced by the Colt Custom Shop between June 1982 and October 1982. Total number delivered 490 in the serial number range 29151s to 29640s.

[2] Produced by Colt Custom Shop in 1976. Factory engraved. Production, 174 units, each built to order and engraved on the backstrap with the owner's name.

CHAPTER ONE: Colt Percussion Pistols

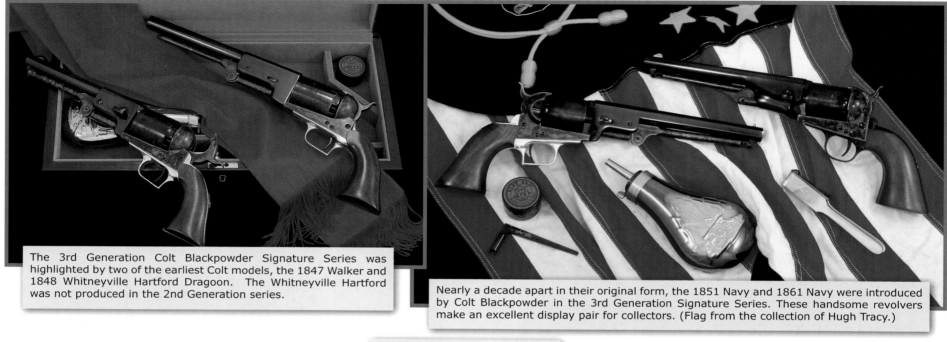

The 3rd Generation Colt Blackpowder Signature Series was highlighted by two of the earliest Colt models, the 1847 Walker and 1848 Whitneyville Hartford Dragoon. The Whitneyville Hartford was not produced in the 2nd Generation series.

Nearly a decade apart in their original form, the 1851 Navy and 1861 Navy were introduced by Colt Blackpowder in the 3rd Generation Signature Series. These handsome revolvers make an excellent display pair for collectors. (Flag from the collection of Hugh Tracy.)

Popular for Civil War reenactments and Cowboy Shooting, the 1862 Pocket Police is one of the more popular 3rd Generation models. Chambered in .36 caliber, the 5-shot revolver has the stylish fluted and rebated cylinder. Reproduction Deputy Marshall badge by Bruce Daly, Mesa, Arizona.

Pockets with a purpose. The 3rd Generation Colt pocket pistols included 1862 Pocket Police (resting in presentation box lid), 1862 Pocket Model of Navy Caliber (in case) 1849 Pocket Dragon, and 1862 Trapper. Guns are pictured with Colt Signature Series accessories.

MODELS FOR A NEW GENERATION OF COLLECTORS

In addition to recreating historic pistols, Colt Blackpowder Arms also established its own contemporary commemoratives in the tradition of the 2nd Generation Ulysses S. Grant, Robert E. Lee set. The first in this new series was the General George Armstrong Custer 1861 Navy. Chambered in .36 caliber, the Custer had an antique silver blue finish egraved in the style of Louis Nimschke, one of the premiere engravers of Colt revolvers in the second half of the 19th century. Presentation stocks for the Custer were select rosewood, engraved with a carved American eagle over a shield on the left-hand side, with a deep checkered pattern appropriate to the era on the right.

While the formal presentation of firearms to American Indians was a rare occurrence, Samuel Colt did deliver his pistols into the hands of a few prominent tribal chieftains. The Colt Blackpowder Signature Series Cochise Dragoon was the first black powder pistol ever produced to honor the great chief of the Chiricahua Apache tribe.

One of the most famous American Indians of the 19th century, Cochise was rarely seen without a large Colt revolver, thus this commemorative was a large-frame 3rd Model Dragoon of the 1851 design, finely wrought with 24-kt. gold inlay depicting a herd of stampeding wild horses along the length of the barrel. The frame was embellished with a 24-kt. gold inlay buffalo, with a gold tomahawk and peace pipe reproduced on the barrel lug to symbolize the roll of Cochise as both a warrior and a peacemaker.

The entire cylinder of this powerful .44 caliber Dragoon was 24-kt. gold-plated, with the battle scene of the Texas Rangers and an Indian war party roll-engraved precisely as it was first executed for Colt in 1847 by master artisan Waterman Lilly Ormsby. The finishing touch to this limited edition Dragoon was a grip set of grained blackhorn engraved with an image of the legendary Indian Chief.

Within the 1860 Army series, two special models were added to the standard rebated cylinder .44, fluted cylinder .44, and nickel plated .44 versions. The first was an 1860 Officer's Model with extremely brilliant blue finish and 24-kt. gold crossed-saber emblem above the wedge. The gun had a fully fluted cylinder and four screw frame cut for a shoulder stock. Unfortunately the stocks were never made available from Colt Blackpowder Arms.

The second special 1860 model introduced in 1996 was the Gold Cavalry, a deluxe version of the 1860 with a 24-kt. gold-plated, rebated cylinder engraved with the crossed saber emblem and 1860 markings in place of the traditional roll engraved battle scene. The frame, loading lever, plunger, and hammer were all color case hardened and 24-kt. gold bands encircled the end of the muzzle. Both the backstrap and trigger guard were plated in 24-kt. gold and protected with a unique patented lacquer.

The last custom Colt Blackpowder reproductions were the Colt Heirloom 1860 Army Tiffany-style revolver, engraved 1842 Paterson No. 5 Holster Model (both of which are covered in Chapter Four) and 1851 Navy 100th Anniversary Commemorative, introduced in 2001. In 2002, the last year of 3rd Generation production, Colt Blackpowder Arms added a standard version of the 1842 Paterson, bringing the total number of Colt 3rd Generation pistols to 28. The greatest number of reproduction models in Colt history and the finest selection ever offered to collectors and shooters.

The big and small of it. Black powder Colts ranged from the massive .44 caliber Walker (center) and clockwise, to 1st, 2nd and 3rd Model Dragoons, Pocket Model of Navy Caliber, Baby Dragoon, 1862 Police, 1861 Navy, 1860 Army rebated and fluted cylinder versions, and 1851 Navy. In addition there was the Whitneyville Hartford Dragoon, 1862 Trapper, 1849 Pocket Dragoon, and Paterson.

Popular during the Gold Rush days, and as a back up pistol during the Civil War, the compact .31 caliber 1848 Pocket Dragoon was introduced by Colt Blackpowder in 1995. This was another model not produced in the 2nd Generation.

Colt Signature Series revolvers featured the flamboyant "Sam Colt" Signature on the backstrap of each model. The 3rd Generation pistols came in a hard sided gray presentation box, while the 2nd Generation Colts came in black and gold soft sided presentation boxes. Saving all original brochures and instruction books is a must for serious collectors.

Colt Blackpowder manufactured a complete line of accessories for each caliber revolver. The kits included powder flask, bullet mold, combination tool, and percussion cap tin. The mold, flask and combination tool all bore the Sam Colt signature.

CHAPTER ONE: Colt Percussion Pistols

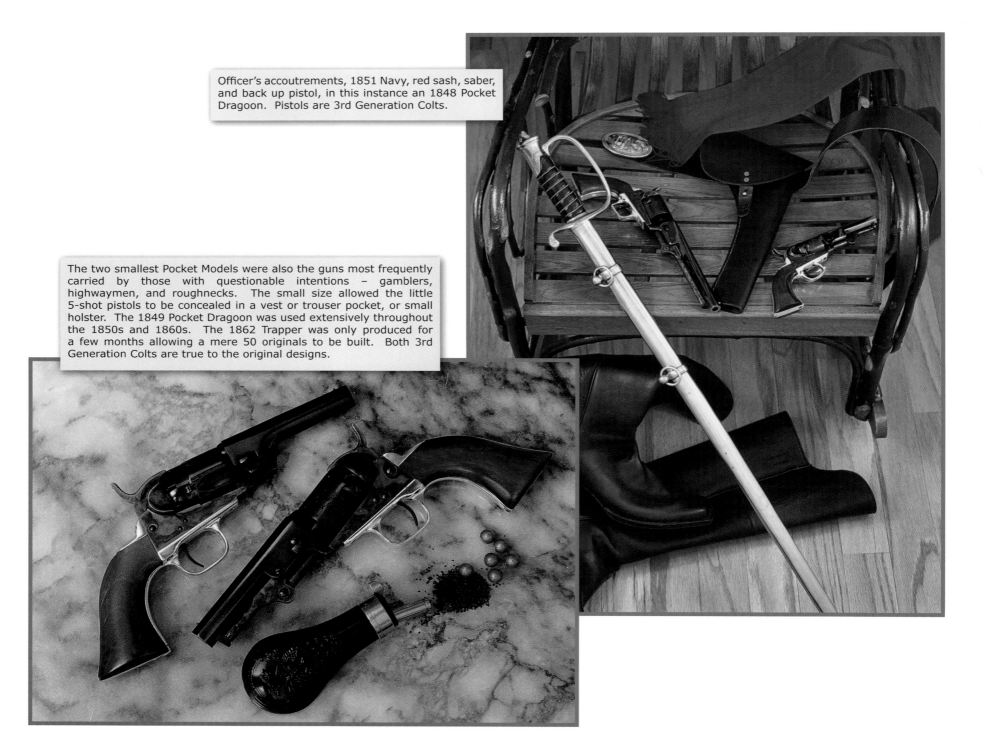

Officer's accoutrements, 1851 Navy, red sash, saber, and back up pistol, in this instance an 1848 Pocket Dragoon. Pistols are 3rd Generation Colts.

The two smallest Pocket Models were also the guns most frequently carried by those with questionable intentions – gamblers, highwaymen, and roughnecks. The small size allowed the little 5-shot pistols to be concealed in a vest or trouser pocket, or small holster. The 1849 Pocket Dragoon was used extensively throughout the 1850s and 1860s. The 1862 Trapper was only produced for a few months allowing a mere 50 originals to be built. Both 3rd Generation Colts are true to the original designs.

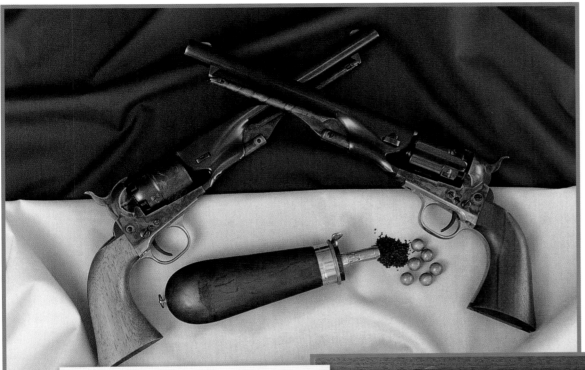

In both the 2nd and 3rd Generation, Colt produced the 1860 Army with rebated and fluted cylinders. Used by both the Union Army and Confederate forces (the latter with captured guns) the 1860 models were the most commonly used sidearms in the Civil War.

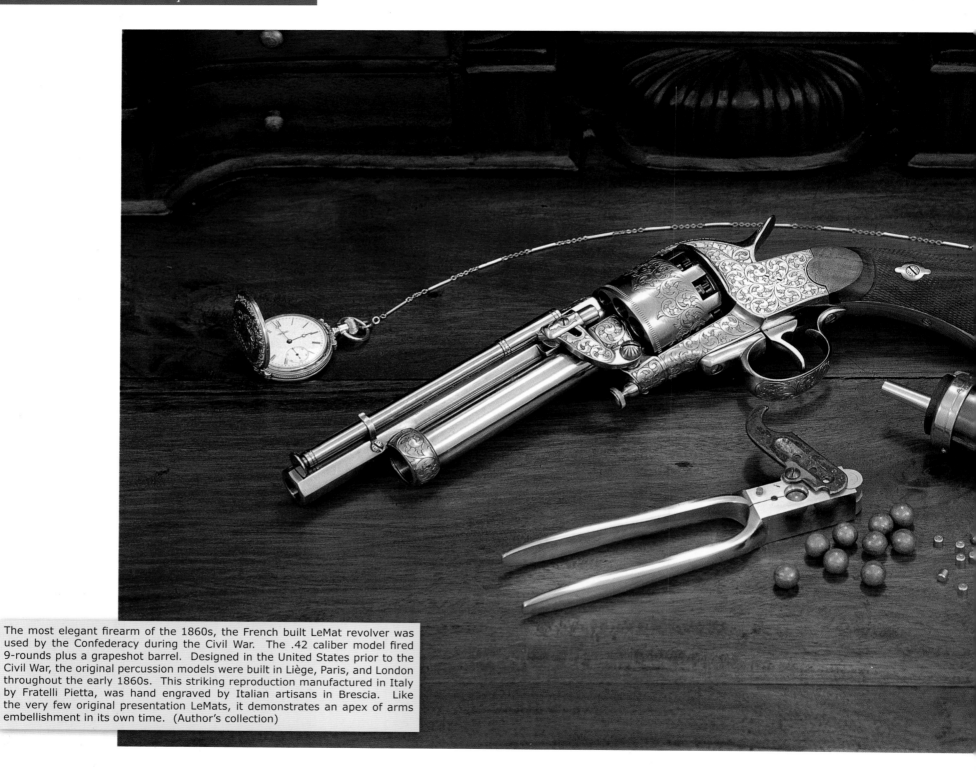

The most elegant firearm of the 1860s, the French built LeMat revolver was used by the Confederacy during the Civil War. The .42 caliber model fired 9-rounds plus a grapeshot barrel. Designed in the United States prior to the Civil War, the original percussion models were built in Liège, Paris, and London throughout the early 1860s. This striking reproduction manufactured in Italy by Fratelli Pietta, was hand engraved by Italian artisans in Brescia. Like the very few original presentation LeMats, it demonstrates an apex of arms embellishment in its own time. (Author's collection)

Colt's Contemporaries 2

Remington, Rogers & Spencer, Spiller & Burr, Starr, Dance, Griswold & Gunnison, and LeMat

Samuel Colt had more than a little competition when he perfected his 5-shot revolver in 1836. Contemporaries of Colt in the field of revolving chambered arms included J. Miller (U.S.A.; patent of June 1829), Samuel Faries (patent of October 10, 1829), David G. Colburn (U.S.A.; patent of October 25, 1832), Isaac Dodds (England; English patent #6826, April 13, 1835), Otis W. Whittier (U.S.A.; patent #216, May 30, 1837), E.A. Bennett and F.B. Haviland (U.S.A.; patent #603, February 15, 1838), Mighill Nutting (U.S.A.; patent #713, April 25, 1838), Elijah Jacquith (U.S.A.; patent #832, July 12, 1838), E.B. Butterfield (U.S.A.; patent of March 16, 1839), and David Edwards (U.S.A.; patent #1134, April 27, 1839),[1] along with many other would-be gunmakers who had designs on a multiple shot repeater. Colt, however, had what proved to be one of the most practical, and a patent of 1836 (aided by a later extension) which would prevent anyone in the United States from legally producing a superior revolver of single action type until the year of expiration, 1857.[2]

The demise of the Patent Arms Mfg. Co. and the years which passed between production of the last Colt Paterson and the prototype Walker, allowed a number of competitors, primarily in Europe, to enter the percussion pistol field. Colt, however, certainly prevailed over all of his contemporaries when the .44 caliber Walker Dragoon was introduced in 1847.

Colt's greatest American competitor was Remington. The manufacture of firearms by Remington dated back to 1816, when Eliphalet Remington produced his first guns in New York.

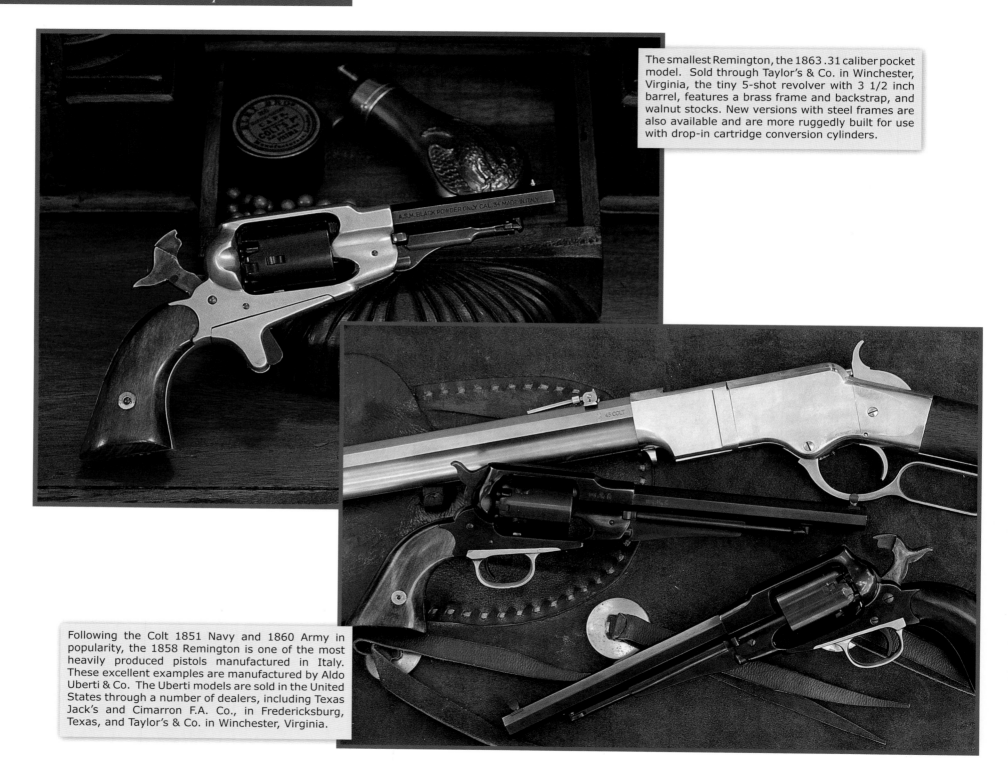

The smallest Remington, the 1863 .31 caliber pocket model. Sold through Taylor's & Co. in Winchester, Virginia, the tiny 5-shot revolver with 3 1/2 inch barrel, features a brass frame and backstrap, and walnut stocks. New versions with steel frames are also available and are more ruggedly built for use with drop-in cartridge conversion cylinders.

Following the Colt 1851 Navy and 1860 Army in popularity, the 1858 Remington is one of the most heavily produced pistols manufactured in Italy. These excellent examples are manufactured by Aldo Uberti & Co. The Uberti models are sold in the United States through a number of dealers, including Texas Jack's and Cimarron F.A. Co., in Fredericksburg, Texas, and Taylor's & Co. in Winchester, Virginia.

It was Remington's 1858 Army model, however, which contested for Colt's title almost immediately after his patent expired.

The Remington was in many respects a superior firearm. With a fixed barrel, take down for cleaning or cylinder exchange was simply a process of half cocking the hammer, lowering the loading lever and removing the cylinder pin. It was also far safer to carry with a notched hammer rest located between each chamber. Colt also added a hammer rest on later models but the Remington design was better.

The hefty Remingtons were chambered for .44 caliber and a second version, the 1858 New Model Navy, was produced in .36 caliber as a direct competitor to the Colt 1851 Navy.

With the beginning of the Civil War, the Union Army placed a substantial order for both .44 and .36 caliber Remingtons, particularly the New Model Army .44 which became the second most carried sidearm of the war.

In need of as many revolvers as possible for the cavalry, infantry, and navy, in 1862 President Abraham Lincoln commissioned Marcellus Hartly, a partner in the New York firearms importing firm of Schyler, Hartly & Graham, to supply the Union Army with French Lefaucheux (pronounced lue-foe-sho) pistols and pinfire ammunition. The early cartridge-firing Lefaucheux was the fourth most commonly used revolver in the American Civil War, surpassed only by the Colt, Remington, and Starr percussion pistols.

It came as no surprise to the Confederacy that with Colt's entire production allotted to the Union Army, the South would have little choice but to build its own version of the Colt repeaters. As such, most were based on the 1851 Navy, a gun most Southerners already owned.

In 1862 the government of Jefferson Davis and the Confederate States of America appealed to the patriotism of anyone who could contribute in the production of guns. Among a handful that came to the aid of the Confederacy was Samuel Griswold, owner of a large cotton mill. Working with A. Gunnison, he converted the mill into a factory and founded Griswold and Gunnison Company in 1862. The firm immediately contracted with the new Southern government in Richmond, Virginia, to take its entire production. Using the 1851 Navy as a model, the general design of the Colt was copied, however, with none of the fine attention to detail that had been the hallmark of the Hartford percussion pistols. The G&G cylinder was a rebated design copied from the Colt 1860 Army, and the frame, made of brass, featured a round trigger guard.

Griswold and Gunnison, like many arms makers during the war, was plagued with material shortages and appealed for help. The Confederacy asked that churches donate their steeple bells for arsenal purposes. The bells allowed the company and other gunmakers to continue production in the early years of the war.

In the winter of 1861-62, J.H. Dance and Brothers, located east of Columbia (first capital of Texas), and 10 miles away from Angleton, Texas, also started production of revolvers. The prototype was presented on April 22, 1862 and a few months later the .36 caliber Dance was in production. Again the design was based on the Colt 1851 Navy, but Dance further

Uberti produces the 1858 Remington in both blued steel and stainless steel finishes. The Ubertis are considered among the best Remington black powder reproductions built for fit, finish, and reliability.

CHAPTER TWO: Colt's Contemporaries

LeMat reproductions were slightly modified by Val Forgett, Sr. and Fratelli Pietta to prevent guns from being artificially aged and passed off as originals. The most significant change is the caliber, altered from .42 to .457, the same standard as other .44 caliber percussion reproductions. In his book on LeMat, Forgett recommends either .27 grapeshot, (around a dozen backed by 35 grains of powder), or a .62 caliber round ball. Pictured with a double gun presentation case from Navy Arms, an engraved Army model, 18th Georgia LeMat Navy, and Beauregard LeMat Cavalry (with trigger guard spur and lanyard ring).

Handsomely engraved Army and Navy versions of the LeMat can be special ordered. The engraving style is in the motif of the Civil War era and with their polished stainless steel finish the engraved guns are durable enough to be fired and cleaned without guilt.

simplified the revolver by eliminating the recoil shield, thus creating a flat sided frame. The Navy's octagonal barrel was also replaced by a simpler half round half octagonal design similar to that of the 1848 Colt Dragoon. Following the same lines, the Dance also used a square back trigger guard.

Among the handful of models produced in the South during the war, the Dance was considered one of the finest arms of the period. The guns remained popular following the war, and among famous owners were Apache warrior Geronimo, and infamous highwayman Bill Longley, who is said to have killed his first man with an 1862 Dance.

A true hybrid design, the 1862 Spiller & Burr was a combination of the Remington solid frame with a Colt mechanism. The gun was designed by Edward Spiller, nephew of James Henry Burton, lieutenant colonel of the Army, and David Burr, a respected Richmond, Virginia, industrial engineer. Together they formed Spiller & Burr, receiving a contact to produce 15,000 revolvers in a period of two and a half years from the date of the order. A production facility was set up in a government-owned factory in Atlanta. From the very start, the company fell well behind schedule, essentially violating the terms of their contract. Spiller and Burr were forced to sell the business to the Confederate Government, which moved production to Macon, Georgia, for the duration of the war.

By far the most famous sidearm used by the Confederacy was the LeMat, perhaps the only percussion pistol ever designed that rivaled the 1847 Walker Colt for its sheer audacity.

The LeMat was a unique firearm, bold in its design, and more so in its presentation, a 9-shot .42 caliber revolver with a secondary lower "shotgun" barrel chambered for grapeshot – the coup de grace.

Dr. Jean Alexandre Francois LeMat left his homeland in France and moved to the United States in 1843 to study medicine. After establishing himself in New Orleans, he married a young American girl, Justine Sophie LePretre, in 1849. She was the cousin of Pierre-Gustave Toutant Beauregard, a U.S. Army major who would later lead the bombardment of Fort Sumter in Charleston Harbor in 1861, and go on to become one of the top generals of the Confederate Army.

A practicing physician in New Orleans, LeMat was also an avid inventor, and Beauregard encouraged and even financed some of LeMat's daring ideas in such diverse fields as medicine, navigation, and firearms. Together with Beauregard, he developed the massive 9-shot revolver which would be carried by many of the Confederacy's top generals, including Beauregard and J.E.B. Stuart, by officers of the Confederate States Navy, and the 18th Georgia Regiment.

The gun had been developed by LeMat prior to the Civil War and was ready for production in 1859 and the very first examples were manufactured in Philadelphia by gunmaker John H. Krider. Dr. LeMat's inventions, including the tracheal speculum, an instrument for spreading and opening the trachea (the airway to the lungs) during surgery, along with the ingenious revolver (quite a contrast!) brought him to the attention of the Governor of Louisiana, who commissioned LeMat a colonel and appointed him his aide-de-camp on April 16, 1859.

LeMat went into partnership with Charles Frederic Girard, Assistant Secretary at the Smithsonian Institution, in July 1860. Like LeMat, Girard had also immigrated to the United States from France. A firearms designer as well, he had received a French patent in 1858 for a breech-loading rifle of his own design. Girard acted as LeMat's European agent and after the beginning of the Civil War he returned to France where he established Girard & Cie., which handled production of the LeMat Revolvers in Liège, Paris, and London between 1862 and 1865. Approximately 3,000 LeMat revolvers were produced before the company went bankrupt following the collapse of the Confederacy. [3]

Three general versions were produced for the Confederate military: the Cavalry model with spur trigger guard, lever-type barrel release, cross pin barrel selector (primary 9-shot cylinder or lower grapeshot barrel), and a lanyard ring; the Navy model with knurled pin barrel release and spur barrel selector; and Army model, which was similar to the Navy but fitted with the cross pin barrel selector. In addition some 100 Baby LeMats chambered in .32 and .41 caliber were produced as well as early French Lefaucheux pinfire cartridge LeMats. There were even a handful of LeMat carbines produced.

The LeMat was probably the most impressive looking revolver of the 1860s although certainly a handful to cock, unless you had thumbs like a

In 1998, Fratelli Pietta in Brescia, Italy, introduced its fully authentic reproductions of the Starr single and double action .44 caliber revolvers used during the Civil War. The double action models were copied in exact detail from the original (un-blued) example in the photo. The top break design (center image) gave the Starr a decided advantage for reloading by changing out cylinders.

lumberjack. Today, original specimens are highly prized among collectors, and the reproductions, handcrafted by Pietta, are ranked among the finest percussion revolvers made.

Once again Clint Eastwood can take the credit for bringing an otherwise obscure revolver to notoriety among gun enthusiasts; the Starr double-action percussion revolver was featured in the Academy Award winning film *Unforgiven*. The .44 caliber Starr was second only to the Colt 1860 Army and Remington Army revolvers among U.S. troops. The Federal Government purchased 47,952 Starr double and single action revolvers from the New York arms maker during the Civil War.

Eben Starr was literally years ahead of everyone with a double action top break revolver. The first Starr revolvers were chambered in .36 caliber. By 1862 demand from the War Department brought about the addition of a .44 caliber version, which amounted to 16,100 guns by May of 1863. A single action version with a longer 8 inch barrel was also requested (the double action models had 6 inch barrels) and this became the most prolific of the Civil War era Starr arms, with production of the 1863 single action model reaching 25,000 by the end of 1864.

The advantage of the Starr was the top break design. Although Eben Starr didn't patent the idea (Sam Colt had actually proposed such a design among his 1850 Dragoon patents), Starr made improvements to the concept by mortising the top strap to fit over the standing breech, thus giving his guns incredible strength. Starr revolvers were built to withstand the punishment of heavy use, yet by simply unscrewing the large knurled cross bolt that passed through the breech, the barrel and topstrap separated from the frame and pivoted down allowing an empty cylinder to be replaced in a matter of seconds. The gun could also be loaded conventionally using the rammer and plunger.

The Starr's unique cylinder design did away with the conventional center arbor (around which both Colt and Remington cylinders revolved), and instead the long ratchet shaft seated into the breech at the rear and locked into the front of the frame with a conical bolt extending from the cylinder.

The Starr double action percussion revolver, like so many others, came to attention when it was featured in a Clint Eastwood western. The Starr was used by Eastwood in his Academy Award winning film, *Unforgiven*.

CHAPTER TWO: Colt's Contemporaries

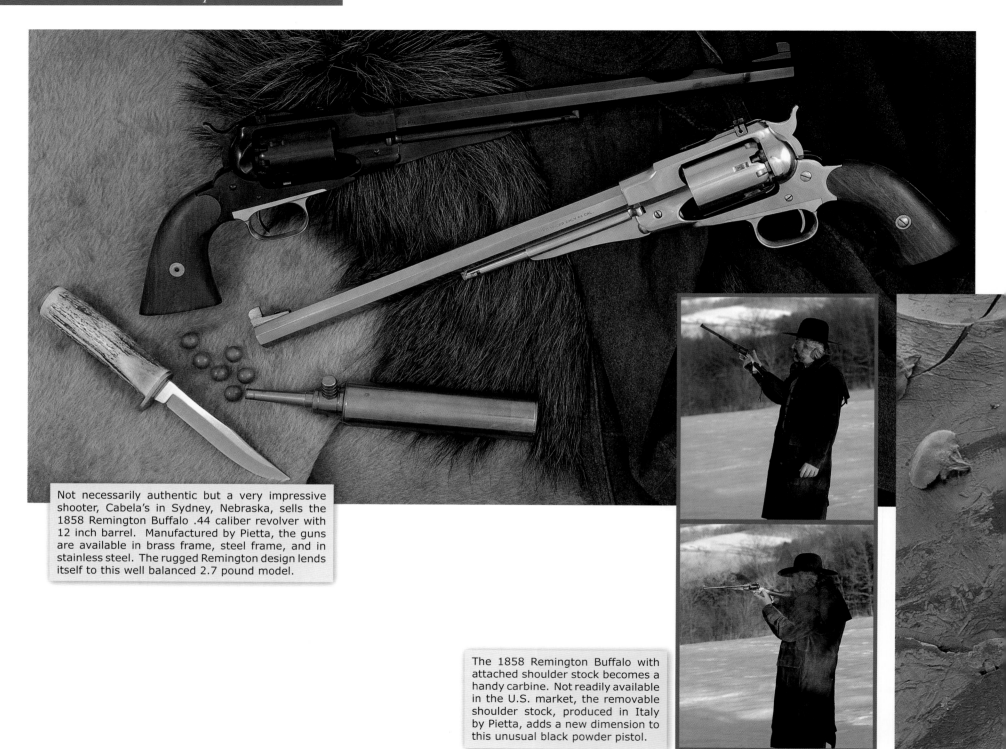

Not necessarily authentic but a very impressive shooter, Cabela's in Sydney, Nebraska, sells the 1858 Remington Buffalo .44 caliber revolver with 12 inch barrel. Manufactured by Pietta, the guns are available in brass frame, steel frame, and in stainless steel. The rugged Remington design lends itself to this well balanced 2.7 pound model.

The 1858 Remington Buffalo with attached shoulder stock becomes a handy carbine. Not readily available in the U.S. market, the removable shoulder stock, produced in Italy by Pietta, adds a new dimension to this unusual black powder pistol.

Because of this design Starr revolvers were less apt to foul since they did not have to rotate around an arbor. As good a design as the Starr was, without government contracts the New York firm found that it could not be competitive with Colt and Remington in the civilian market, and two years after the war, Ebenezer Townsend Starr ceased production of the guns, revolvers that had perhaps been too advanced for their time.

Another late comer was the Rogers and Spencer .44 which was introduced toward the very end of the Civil War and barely in use by the time hostilities ceased in May, 1865. The large frame revolvers resembled Remingtons, but were considerably different in execution, based on the Freeman revolver engineered by Austin T. Freeman, who had been an employee of the Starr Armory in New York.

The Freeman revolvers in .44 caliber were originally produced by C.B. Hoard in Watertown, New York, but became a product of Utica, New York, firearms manufacturer Rogers and Spencer late in 1864. The R&S model combined the Freeman design with that of a Remington, and the double action Pettengill revolver which Rogers and Spencer had previously produced. The guns also resembled the Whitney revolver with the overall design of the Rogers and Spencer combining the front frame of a Remington with the back frame of a Starr, Starr-type loading lever, and Pettengill backstrap and stocks.

Ordered as a single action pistol, 5,000 new .44 caliber Rogers and Spencer revolvers were requested by the Ordnance Department in January 1865, but the order wasn't completed until September, four months after

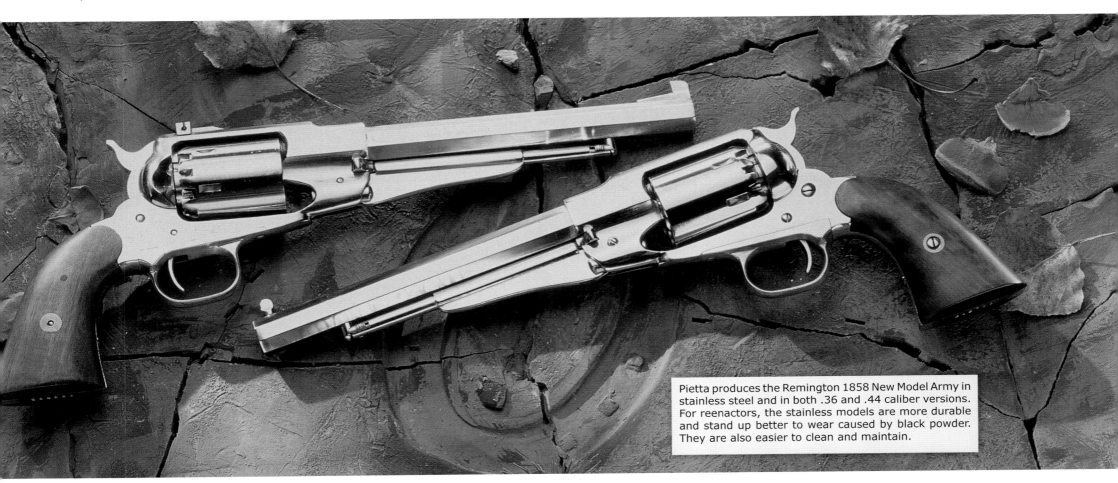

Pietta produces the Remington 1858 New Model Army in stainless steel and in both .36 and .44 caliber versions. For reenactors, the stainless models are more durable and stand up better to wear caused by black powder. They are also easier to clean and maintain.

the last Confederate Army had surrendered. Most of the pistols delivered to the Ordnance Department remained in storage in New York until 1904!

The Rogers and Spencer was an exceptionally strong pistol, although it had what could only be described as a long, hard hammer draw. Aside from that, it was a rugged, reliable sidearm that saw later service in the Spanish American War.

All of the great Civil War era models produced by Remington, Dance, Spiller and Burr, Starr, Griswold and Gunnison, Rogers and Spencer, and LeMat, have been faithfully reproduced in Italy by F.lli Pietta, with the exception of the Rogers and Spencer revolver, which is produced by Euroarms of America (manufactured in Italy) and distributed along with all of the Pietta models through Dixie Gun Works in Union City, Tennessee.

The quality of Pietta reproductions (discussed in Chapter Six) and that of Uberti and several of the Brescia and Gardone Valley gunmakers, is well beyond that of the original 19th century models. It is rewarding to collectors and enthusiasts of these historic firearms to have such a wide variety of authentic reproductions available, particularly models as highly complex in design as the LeMat and Starr, the latter being made available in 1998 after more than two years of development by F.lli Pietta.

These quality firearms add immeasurably to preserving the memory of early America and the Old West, providing a sense of our great heritage that would otherwise have been lost to the ravages of time, and left only to the pages of books, and antiques displayed in collections and museums.

The popular Remington design produced by Pietta is also offered in steel frame, and in a brass frame Confederate model. The Buffalo with brass frame is the least expensive of the long barreled custom models.

Black Powder Revolvers - Reproductions & Replicas

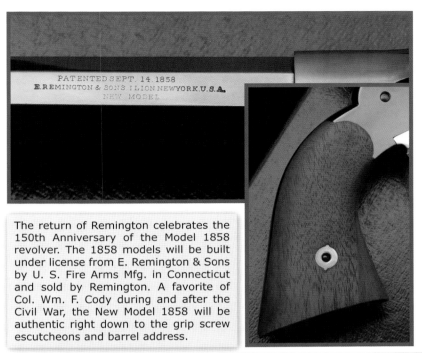

The return of Remington celebrates the 150th Anniversary of the Model 1858 revolver. The 1858 models will be built under license from E. Remington & Sons by U. S. Fire Arms Mfg. in Connecticut and sold by Remington. A favorite of Col. Wm. F. Cody during and after the Civil War, the New Model 1858 will be authentic right down to the grip screw escutcheons and barrel address.

The South may not rise again, but the guns of the Southern Confederacy are doing quite well. The entire Reb line from Pietta includes the 1860 Griswold and Gunnison cased model; 1862 Spiller and Burr, (left front); 1851 Reb short, (left rear); 1862 Dance, (right front); and 18th Georgia LeMat.

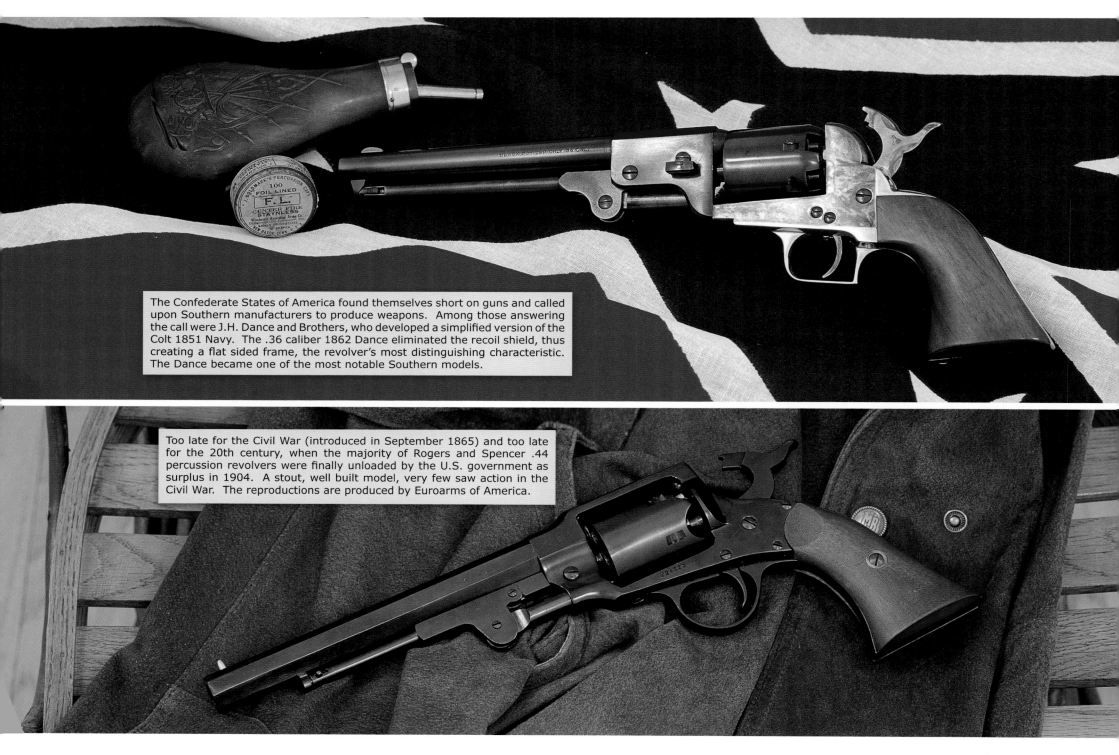

The Confederate States of America found themselves short on guns and called upon Southern manufacturers to produce weapons. Among those answering the call were J.H. Dance and Brothers, who developed a simplified version of the Colt 1851 Navy. The .36 caliber 1862 Dance eliminated the recoil shield, thus creating a flat sided frame, the revolver's most distinguishing characteristic. The Dance became one of the most notable Southern models.

Too late for the Civil War (introduced in September 1865) and too late for the 20th century, when the majority of Rogers and Spencer .44 percussion revolvers were finally unloaded by the U.S. government as surplus in 1904. A stout, well built model, very few saw action in the Civil War. The reproductions are produced by Euroarms of America.

Black Powder Revolvers - Reproductions & Replicas | 51

For 2008 America Remembers will pay tribute to the great Apache leader Geronimo with a limited edition Dance Dragoon engraved in an authentic Plains Indian motif with brass tacks in the grips, a silver rosette, Apache pictograph paintings, and 24-karet gold busts of Geronimo and an American Bison. The backstrap has the symbol of the medicine man and Geronimo's name. The gun was designed by Andrew Bourbon, John J. Adams, Sr., and the author. The edition will be limited to 300 guns set into gilt-edged book case presentation boxes.

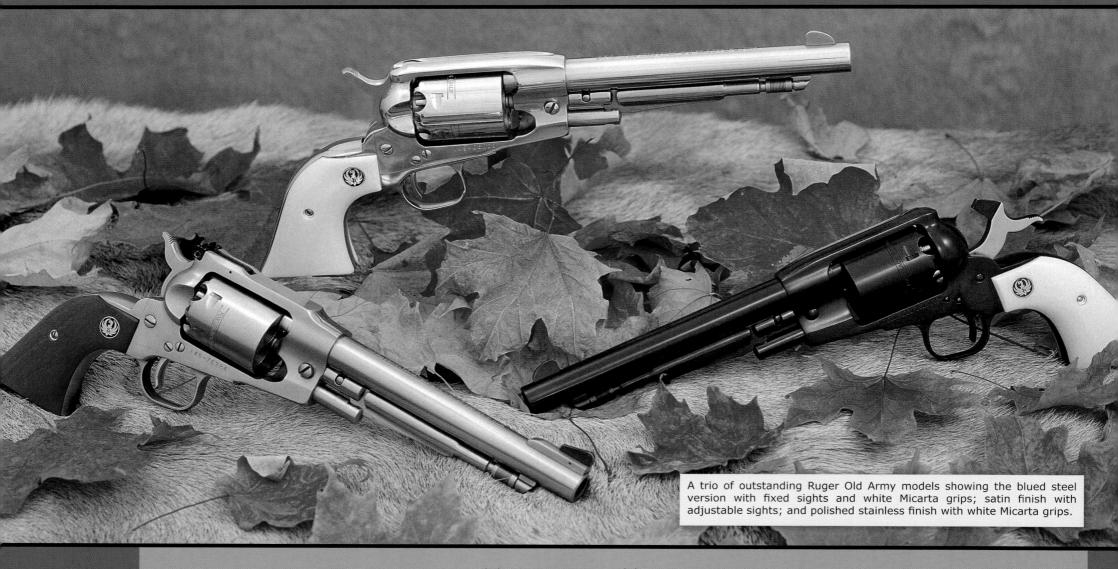

A trio of outstanding Ruger Old Army models showing the blued steel version with fixed sights and white Micarta grips; satin finish with adjustable sights; and polished stainless finish with white Micarta grips.

THE RUGER .44 OLD ARMY

The best-built 19th century percussion revolver of the 21st Century

I had the pleasure of knowing William B. Ruger, Sr. for many years, as well as Bill Ruger, Jr., both through my interest in firearms and theirs in classic American and European cars. Our shared love of old automobiles and old guns has made me an avid Ruger Old Army enthusiast, and after firing countless Colt, Remington, and LeMat reproductions for this book, I can irrefutably attest to the fact that there is no better shooting black powder revolver made today than the Old Army. And for the record, had I never become acquainted with the Ruger family I would still feel the same way

Introduced in 2003, the 5 1/2 inch Old Army has become one of the most popular cap and ball guns in SASS black powder competition. It was available in blued or polished stainless finish with wood or white Micarta grips.

about these superbly built percussion handguns. In fact, for the company's 50th Anniversary I joined with Bill Ruger, Jr. in designing a special anniversary model which later became the now popular 5 1/2 inch model Old Army.

As a firearms collector, Wm. B. Ruger always had a keen interest in black powder guns, and had said years before the Old Army was developed that, "It would be nice to make a percussion revolver that was a really good shooter, and as close to indestructible that could be made, with all the usual features."[4]

Although Sturm, Ruger & Co. has been a technological innovator for over half a century, the majority of the company's pistols, rifles and shotguns have been inspired by classic firearms of the American West. Bill once told me, "There is no one who can design a firearm without reference, without some connection to an earlier design. Just as engines, fuels, and aerodynamics help define what an automobile will look like, the dynamics of a bullet and how it works, defines a firearm."

As a young firearms designer for Auto Ordinance in the 1940s, Ruger was introduced to the secrets of investment casting. Also known as the lost wax process, investment casting has been used since the time of the Pharaohs to create jewelry, sculpture and tools, and in the automotive industry to manufacture everything from an engine block, or a door handle, to a Rolls-Royce hood ornament. Of course, there's quite a difference between an engine block and the inner workings of a pistol, enough, that it took Ruger more than a decade after he introduced his first pistol in 1949 to advance the casting technology to its present state, one that can allow even the most intricate pieces to be reproduced to exacting tolerances. "Collectors have come to recognize this trait in Ruger pistols, rifles and shotguns," he told me in 1997. "There is no Model B. Everything is a Rolls-Royce." The Old Army .44 is that, and more.

The rugged Ruger .44 caliber Old Army in blued steel with white Micarta grips is a personal favorite of Bill Ruger, Jr. As a safety feature, the Ruger cylinder has notched hammer rests between each chamber to prevent accidental discharge when the gun is being carried. Tom Mix Stetson courtesy of Hatco.

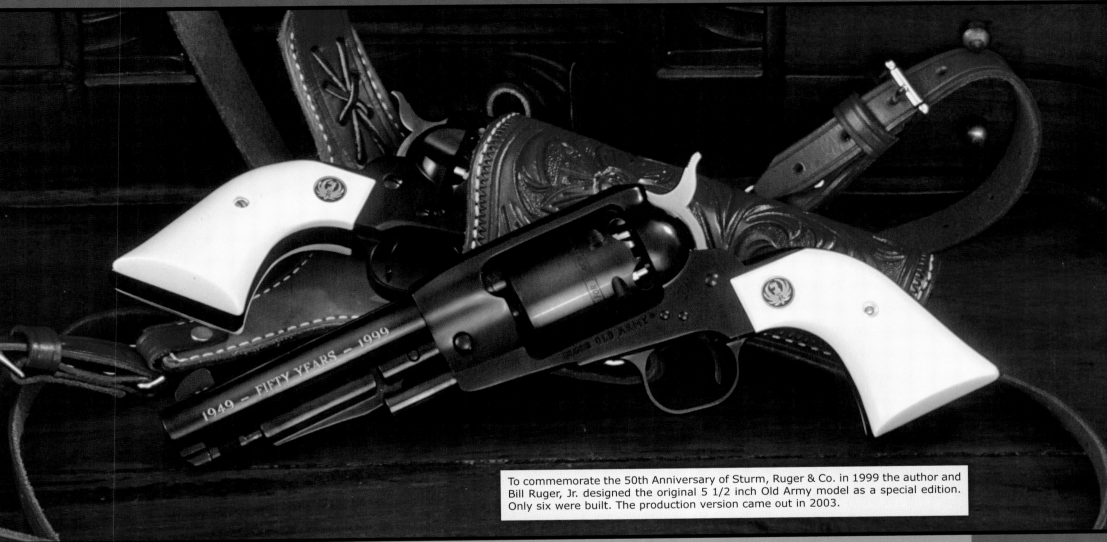

To commemorate the 50th Anniversary of Sturm, Ruger & Co. in 1999 the author and Bill Ruger, Jr. designed the original 5 1/2 inch Old Army model as a special edition. Only six were built. The production version came out in 2003.

In creating the .44 Old Army, Ruger started with a clean sheet of paper. Noted design engineer Harry Sefried in R.L. Wilson's book *Ruger & His Guns*, "None of this stuff of no topstrap, like the Colt percussion revolvers, or simply duplicating an antique design. This sturdy revolver would be basically a Super Blackhawk in percussion, to utilize Blackhawk components as much as possible – grips, gripstraps, etc. One of the details we wanted was no screws (since, for one thing, they always shot loose in percussion revolvers of the 19th century), except for attaching the grip frame and securing the hammer, trigger, and cylinder stop within the frame. I worked that linkage, using the rammer and base pin to secure the cylinder, a coil mainspring system, and a number of other innovations. The fired gun could be cleaned up easily, especially when we came out with a stainless steel version. We wanted a nice big hammer for positive ignition and we even investment cast the cylinder as well. In other words, we combined all the features for which investment castings were perfect, on a frame natural for the .44 caliber. We called the caliber .44, but it's technically a .457, since we used our .45 caliber revolver barrels."

The gun was of incredible strength, as Bill Ruger had insisted it be, but the trial by fire, so to speak, proved just how strong the Old Army was. The test firing of the revolver was conducted at Sturm, Ruger & Co., using Bullseye smokeless powder, "Definitely not recommended to the public!", notes Sefried. Even with the cylinder filled, the Old Army could not be blown up! (Again, not to be tried at home! The only powder suitable for the Old Army is black powder, Pyrodex, or other black powder substitutes).

Utilizing a heavy investment cast frame (similar in appearance to the 1858 Remington) the Old Army came with a 7 1/2 inch barrel, spring-locked loading lever, adjustable rear sight, and 1/8 inch wide flattop Baughman type ramp front sight (changed after serial number approximately 140-00800 to pointed profile) with blue steel blade. In 1994, traditional fixed sights were also made available.[5] Borrowing another design from the 1858 Remington, the Ruger utilized a grooved hammer rest between each chamber to ensure safe carry of the pistol. Considering the Old Army's size (13 1/2 inches) and heavy construction (chrome-molybdenum steel, or stainless steel), it is a surprising 2 pounds.

Introduced in 1972, the Ruger Old Army Revolver quickly earned a reputation for quality and unmatched accuracy. The author's nickel Ruger Old Army has turned in 1 1/2 inch groups at 50 feet, loaded with a conservative 40 grains of Pyrodex powder delivering a Speer .454 round ball. The recommended maximum load is equal to that of a Colt Walker Dragoon, 60 grains of Goex FFFg black powder.

THE OLD ARMY LINE

Major variations to the Old Army included a change in markings beginning with serial number range 140-34882 and 145-08142 in 1978, and continuing to the present. All barrel tops engraved with a warning:

Before using gun-read warnings in instruction manual available free— from Sturm, Ruger & Co., Inc. Southport, Conn. U.S.A.—

Easier to take down and clean than a Colt, the Ruger disassembles by rotating the lock screw in the frame, lowering and removing the loading lever and center pin, allowing the cylinder to drop out of the frame. The quick take down also facilitates easier field cleaning of powder residue to extend shooting time or remove a jammed percussion cap.

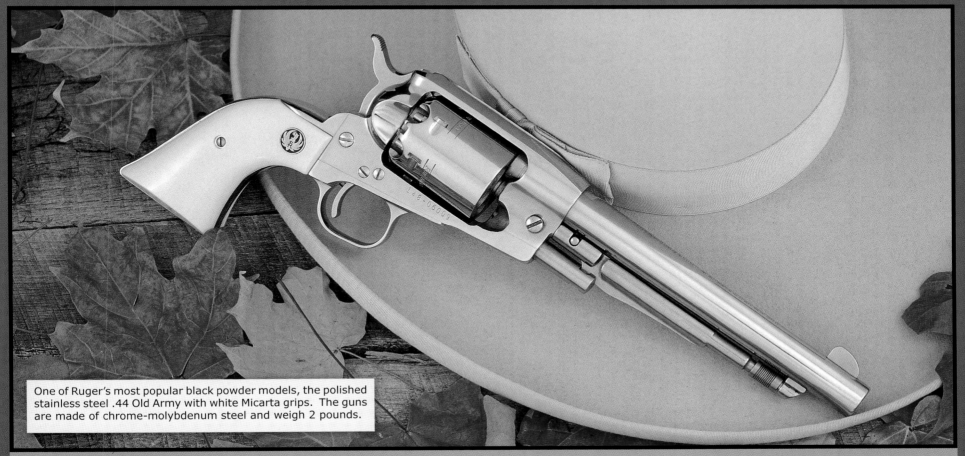

One of Ruger's most popular black powder models, the polished stainless steel .44 Old Army with white Micarta grips. The guns are made of chrome-molybdenum steel and weigh 2 pounds.

The cylinders were also marked twice around the periphery within a border motif: Black Powder Only

The first series of revolvers were all blued and ran from serial number 140-00000 to -46841 (through 1981); serial numbers intermix with stainless steel revolvers from 1982 (with prefix 145-).

The stainless versions were serial numbered 1 to 7790, 145-00000 to -6831 (through 1993; intermixed with blued from 1982.). [6] The total number produced over the past quarter century is well over 150,000. In 2007 Ruger discontiued the Old Army models.

Individual Ruger Old Army models: BP-7: Blued steel, anodized aluminum grip frame; built up through approximately serial number range 140-07700. BP-7B: Brass grip frame, wide trigger. Approximately 1200 blued revolvers built in this variation, serial number range 140-00000 to -04750. BP-7F: Blued steel with fixed sights. (also available with white Micarta grips, BPI-7F). KBP-7: Stainless steel, with adjustable sights. KBP-7B: With brass grip frame. KBP-7F: Stainless steel with fixed sights. GKBPI-7F: Nickel finish with white Micarta grips.

Special Issue Revolvers: National Muzzle Loading Rifle Association: Serial range 140-14000 to -14100. Identified by N.M.L.R.A. logo on the grip panel.

Ruger Collectors Association: Serial range 1500 to 1599, - RCA - marked on the topstrap and a star motif preceding the serial number.

Ruger Collectors Association: Second series, serial range 145-01401

CHAPTER TWO: Colt's Contemporaries

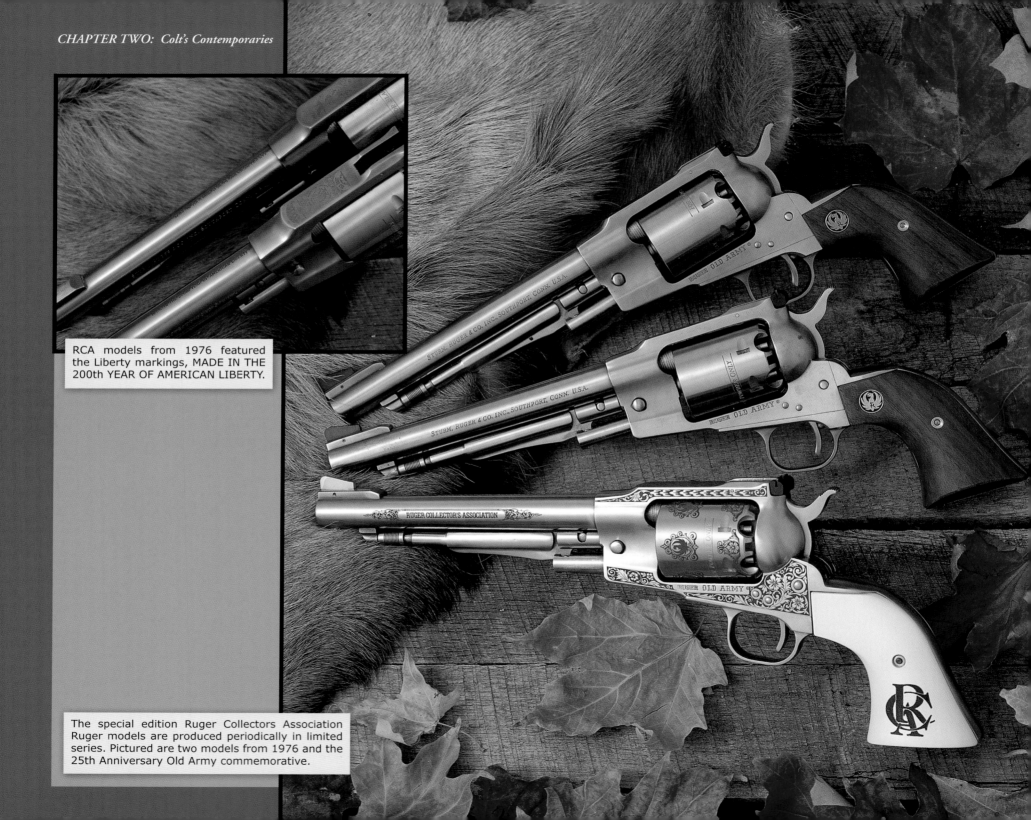

RCA models from 1976 featured the Liberty markings, MADE IN THE 200th YEAR OF AMERICAN LIBERTY.

The special edition Ruger Collectors Association Ruger models are produced periodically in limited series. Pictured are two models from 1976 and the 25th Anniversary Old Army commemorative.

The RCA 25th Anniversary model is the most highly embellished Ruger Old Army ever issued. Only 2000 were made in 1997.

through -01600. Total of 201. Engraved RCA logo and intertwined RCA monogram on topstrap, standard with an eagle marked before the serial number, on grip frame butt, and on right grip panels (some few variations observed, including with U.S. markings). Barrels with Liberty marking, and these 201 revolvers were the last of the 1976 Rugers so marked. (Barrels of Ruger Old Armys produced in 1976 were marked MADE IN THE 200th YEAR OF AMERICAN LIBERTY.)

Ruger Collectors Association: Third series, offered in 1997 to commemorate the 25th anniversary of the Ruger Old Army. Limited to 2000 pistols serial numbers 1 to 2000. Floral engraving on top strap and frame, bottom of trigger guard with RCA logo, Ruger Collectors Association bordered with scrollwork along both sides of the barrel, 24-kt. gold plated cylinder with engraved Ruger emblem and scrollwork, 24-kt. gold plated front sight, white Micarta grips with RCA logo. This is the most elaborate model in the Ruger Collectors Association series.

A total of eight revolvers were embellished with etching, arranged by James M. Triggs. The most elaborate Old Army was a finely engraved stainless steel model with 80 percent coverage in vine scroll motif, serial number 145-48484, done by master engraver Paul Lantuch.

Regardless of the finish, every Ruger Old Army is a rugged, precision-made firearm. The best 19th century percussion revolver of the 21st century.

*****

[1] The Book Of Colt Firearms by R.L. Wilson, 1993 Blue Book Publications. Chapter 1, Samuel Colt And His Early Repeating Firearms, page 4.
[2] Ibid
[3] LeMat text adapted from Valmore J. Forgett and Alain F. & Marie-Antoinette Serpette's book LeMat The Man, The Gun, published in 1996 by Navy Arms.
[4] Ruger & His Guns, by R.L. Wilson, 1996 Simon & Schuster, Chapter V.
[5] Ibid
[6] Ibid

CHAPTER THREE: Reproduction Percussion Pistols

> "Whatever has been made by man can be made again. All it takes is money."
> — Anonymous

That theory has been applied to everything from antique Chippendale desks, Tiffany lamps, and 1934 Packard LeBaron dual cowl phaetons, to 1838 Colt Paterson No. 5 Holster Model revolvers.

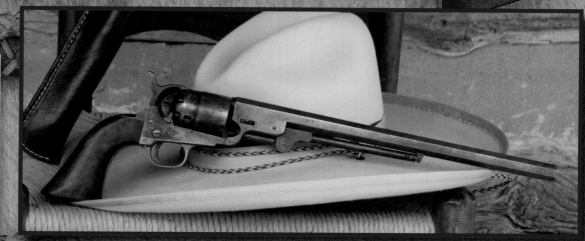

Historically inaccurate but interesting nonetheless, the 1851 Buffalo revolver chambered in .44 caliber looks similar to the very rare, original 1851 Navy models built with 12 inch barrels. The Buffalo features adjustable rear sights with a prominent "U" notch and square cut blade front sight at the far end of a 12 inch barrel. The authenticity issue is the caliber and rebated Army cylinder.

Taking things a bit further R.L. Millington of ArmSport LLC in Platteville, Colorado, built a one-of-a-kind example for the author using a Pietta Army frame cut for a shoulder stock and the 12 inch barrel. An old Pietta shoulder stock was found at VTI Gun Parts in Lakeville, Connecticut, to complete the gun. Jim Lockwood of Legends in Leather in Prescott, Arizona designed a Slim Jim holster rig to accommodate the long barreled Colt.

Reproductions of 19th century Colt black powder revolvers have been popular with gun collectors for more than a quarter of a century. When advanced antique arms collector Bob Hable saw the new percussion line, early in the 1970s, his reaction was to buy one of each, and to store his antiques in a nearby vault – filling up his gunroom display cases with the replicas.

This would be the best of both worlds. But for those whose budgets can't cover several million dollars worth of vintage Colts, or even a full set of 2nd Generation models, there is an abundant supply of high-quality, reasonably priced reproductions available through companies like Navy Arms, Taylor's & Co., Cimarron F.A. Co., and catalog houses such as Cabela's, Bass Pro, and Dixie Gun Works. These

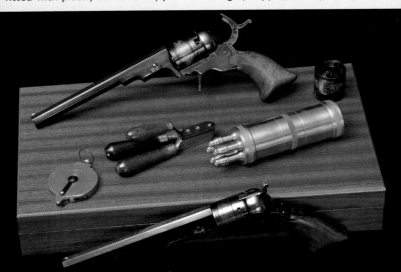

The least produced reproduction Colt is the Paterson No. 5 Holster Model. The guns were primarily manufactured by A. Uberti and F.lli Pietta, and now are only produced in one model, without loading lever, by Pietta. The pair pictured are both Uberti models, a No. 5 with gold inlaid cylinder (special order) and a No. 5 with attached loading lever. The Paterson trigger stows in the frame until the hammer is cocked, at which time it drops down into the ready position. Patersons do not have trigger guards. The guns have been variously marketed individually or in a walnut presentation case French fitted with pistol, bullet mold, powder charger, capper, and spare cylinder.

3

REPRODUCTION PERCUSSION PISTOLS

In The Image Of Col. Colt

CHAPTER THREE: Reproduction Percussion Pistols

Engraved Paterson with etched yellowed Micarta grips was marketed by Texas Jack's and Cimarron F.A. Co. in Fredericksburg, Texas. The highly detailed engraving was done in Italy through A. Uberti which discontinued the Paterson model in 2007.

Two versions of the Texas Paterson were produced by Uberti, the standard No. 5 Holster Model and Holster Model with attached loading lever.

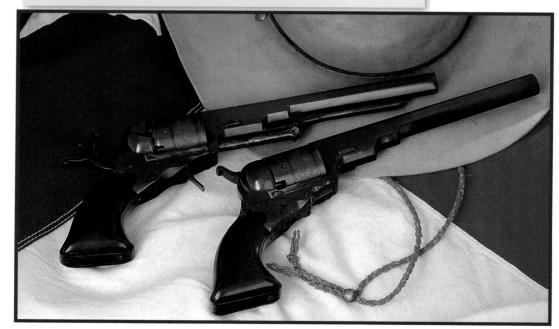

The workhorse of the Old West, Uberti's 1847 Walker is one of the most popular of all black powder reproductions.

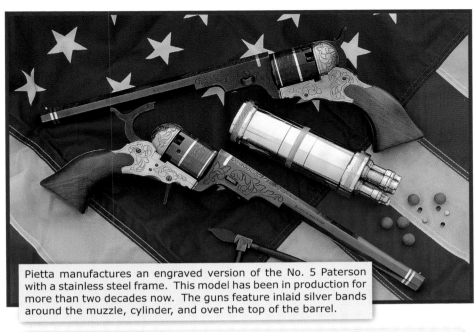

Pietta manufactures an engraved version of the No. 5 Paterson with a stainless steel frame. This model has been in production for more than two decades now. The guns feature inlaid silver bands around the muzzle, cylinder, and over the top of the barrel.

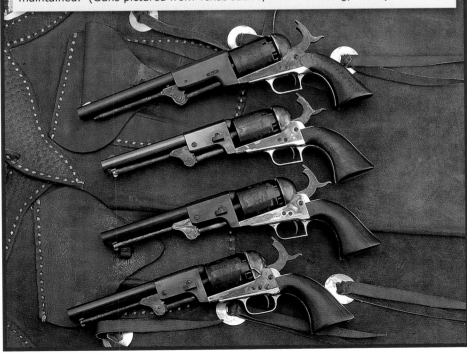

Big guns from Uberti include, from top to bottom, 1847 Walker, 1st, 2nd, and 3rd Model Dragoons. All four models have charcoal blue finish and rugged construction. These are real shooters that will give years of reliable performance when well maintained. (Guns pictured from Texas Jack's, Fredericksburg, Texas)

are the sources for guns produced in the image of Col. Colt's greatest models, everything from Paterson revolvers to pocket pistols.

With the exception of only one gun, the 1862 Trapper, every Colt percussion model is available in a more affordable version produced by the leading gunmakers in Italy – F.lli Pietta and A. Uberti.

WHAT YOU PAY FOR

"What separates the men from the boys is the price of their toys." It is also what separates the quality of reproduction Colt percussion pistols. This comes down to the metallurgy of individual parts (mainsprings and trigger mechanisms in particular), their machining, polishing, fitting, and the case hardening and bluing of each gun. The particulars of how reproduction firearms are produced in Italy is covered in Chapter Six. In this chapter we are going to address fit, finish and quality.

There is no easy way to recreate a 19th century revolver. What is easier is the manufacturing of the individual components that go into a gun. The final assembly, fitting and finish, however, are still a job best done by hand, and that is what separates reproduction revolvers by price.

The feel and sound of an action is an immediate clue to the quality of a firearm. If the hammer draws back smoothly and the progression of clicks as it locks back sound crisp, you probably have a well made pistol. If the action drags and the hammer seems to just stick in place when fully cocked, you have one that could use a visit to the local gunsmith for fine tuning, which may be as simple as deburring and polishing a few internal parts.

The ease with which the cylinder rotates is another crucial point. It should be firm but not stiff, and move in perfect unison with the draw of the hammer. It's a syncopated movement that should sound and feel smooth and crisp. Anything less is a gun that lacks quality and probably has a drag line around the cylinder.

Surface polishing is another distinguishing characteristic; the more preparation prior to bluing or plating, the better the look. Lower-priced guns that have not been finely polished prior to bluing tend to have more porous surfaces, show variations in color, and occasionally have fine underlying scratches in the finish. This is usually found on guns priced at or under $350.

CHAPTER THREE: Reproduction Percussion Pistols

Civil War era Colt sidearms manufactured by Uberti. Shown with an original McClellan cavalry saddle, blanket and saber, are the most widely used pistols in Civil War reenactments, movies and television, the 1851 Navy, 1861 Navy, and 1860 Army with fluted cylinder. (Guns and accessories pictured from Texas Jack's, Fredericksburg, Texas.) An 1861 Navy steel frame model is also available.

Another Uberti sample gun, a Baby Dragoon in silver with full engraving, ivory checkered grips, and gold-plated backstrap and trigger guard.

The Uberti family of standard Colt reproductions, 1840-41 Paterson No. 5 Holster model and No.5 with loading lever (both discontinued in 2007), 1847 Walker, 1848 1st Model Dragoon, 1850 2nd Model Dragoon, 1851 3rd Model Dragoon, 1851 Navy, 1861 Navy, 1860 Army with fluted cylinder, 1860 Army with rebated cylinder, and at top, 1862 Pocket Police and Pocket Navy. All feature Uberti's bright charcoal blue finish and cyanide case hardening, (except Paterson No. 5). (Guns and accessories pictured from Texas Jack's, Fredericksburg, Texas)

A neat custom offered through Texas Jack's is the 1851 Navy James Butler Hickock engraved edition. The gun features white Micarta stocks, vine scroll engraving with punch dot background, and "J.B. Hickock 1869" engraved on the brass backstrap. Wild Bill's last name was correctly spelled Hickok, not Hickock, a common error.

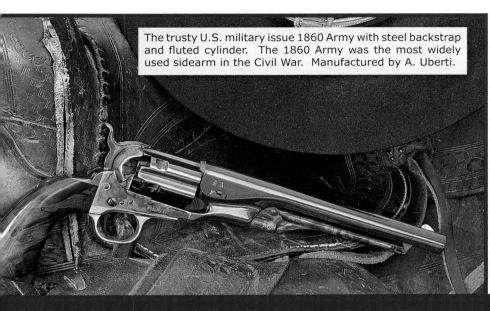

The trusty U.S. military issue 1860 Army with steel backstrap and fluted cylinder. The 1860 Army was the most widely used sidearm in the Civil War. Manufactured by A. Uberti.

Bluing is perhaps the most important feature. Typical Italian bluing has an almost translucent charcoal blue hue, very attractive but lacking the depth of original Colt bluing. (Deeper high-gloss bluing is also done in Italy, but only on premium models, presentation sets, and on engraved pistols).

Case hardening is the last, but by no means least important thing to look for in a quality reproduction. Traditionally, case hardening done in Italy uses a cyanide process which renders a dusky, grey, blue, brown mottling on the surface of the parts treated, whereas true color casehardened parts are done using the centuries old method of heating and cooling the metal, producing brilliant colors. (Case hardening is explained in detail in Chapter Six).

The bottom line, that is, the price you pay, is dependent upon the level of quality one desires. Any black powder revolver, whether it is made by Uberti, Pietta, or another manufacturer will shoot. The difference is in how well it

AMONG THE GOOD, THE BAD AND THE UGLY, IT WAS "THE UGLY" THAT GOT THE MOST ATTENTION

functions and whether it stands up to repeated use without breaking. Thus, like most everything in life, the less you pay, the less you get.

With that said, both Pietta and Uberti occasionally throw the book on fit and finish out the window, and gun owners are clamoring for, the worst looking guns money can buy. Behold the antique or "original" finish, the great reproduction oxymoron, since original should be a new finish and not one that looks like it is 150 years old.

Some call it the patina finish. It's everything you don't look for in a quality replica pistol – scratched, dented, round-edged, pitted, and devoid of 98% of the bluing. "Holster stuffers," generally the kind of gun you would walk right past at an antique gun show. Only these are brand new.

Antique finishes are available on Colt 1851 Navy, 1860 Army and 1858 Remington models from Pietta, and from Uberti in Walker, 2nd Model Dragoon, 1851 and 1861 Navy, and 1860 Army versions.

The Uberti models, made for Cimarron F.A. Co. in Fredericksburg, Texas,

One of the finest examples of Uberti craftsmanship is this engraved 1847 Walker with 24-kt gold inlay. The gun was made for the company's late founder, Aldo Uberti, and is part of the Uberti collection in Italy.

Presentation is almost everything. A double gun case with accessories is a perfect way to display a handsome pair of Uberti 1862 Pockets. Cases are available through Cabela's, Cimarron F.A. Co. and from Taylor's & Co.

Another example of Uberti custom work, this set of 1862 Pocket Police revolvers with full engraving, silver and deep blue finish, and 24-kt gold inlay. These were sample guns made for the Uberti showroom in Brescia by local artisans. Uberti custom work is seldom seen in the United States.

Sam Colt had his own way of delivering justice. This Uberti 1862 Pocket Police "Law and Order" set comes with bullet mold, powder flask, cap tin, and capper, all contained in a replica of the famous French fitted Colt book style casing. The original Colt version was marked along the spine in gold leaf: COLT/ON THE CONSTITUTION/HIGHER LAW & ORDER & IRREPRESSIBLE CONFLICT.

Approximately 129,000 Model 1860 revolvers were issued to U.S. troops for Civil War service, several thousand of them equipped with an attachable shoulder stock, an accessory to allow firing the arm as a carbine. Pietta made a reproduction of this historic model complete with shoulder stock but they are now very hard to find.

have a more distressed finish than do those from Pietta, and the Uberti stocks show greater wear. Cimarron offers the guns under the name "The Original." The Pietta models are sold under the name "Old West."

Both the Pietta and Uberti pistols have the same general appearance – which is beat, really beat – however, the two top Italian gunmakers reached the same end via totally different paths.

Alessandro Pietta explains the technique used to create their patina finish. "After several attempts to use chemicals, we came across a natural oil that could strip the blued finish off a new gun in three to four hours. You then wash away the oil and polish the gun to complete the aging process." That's the short version minus the name of the specific oil used and the finishing details that blemish the gun's denuded surface. The result is a rather handsome, if somewhat worn looking pistol that has what can best be described as "character."

The technique used by Uberti is quite different and produced a far

CHAPTER THREE: Reproduction Percussion Pistols

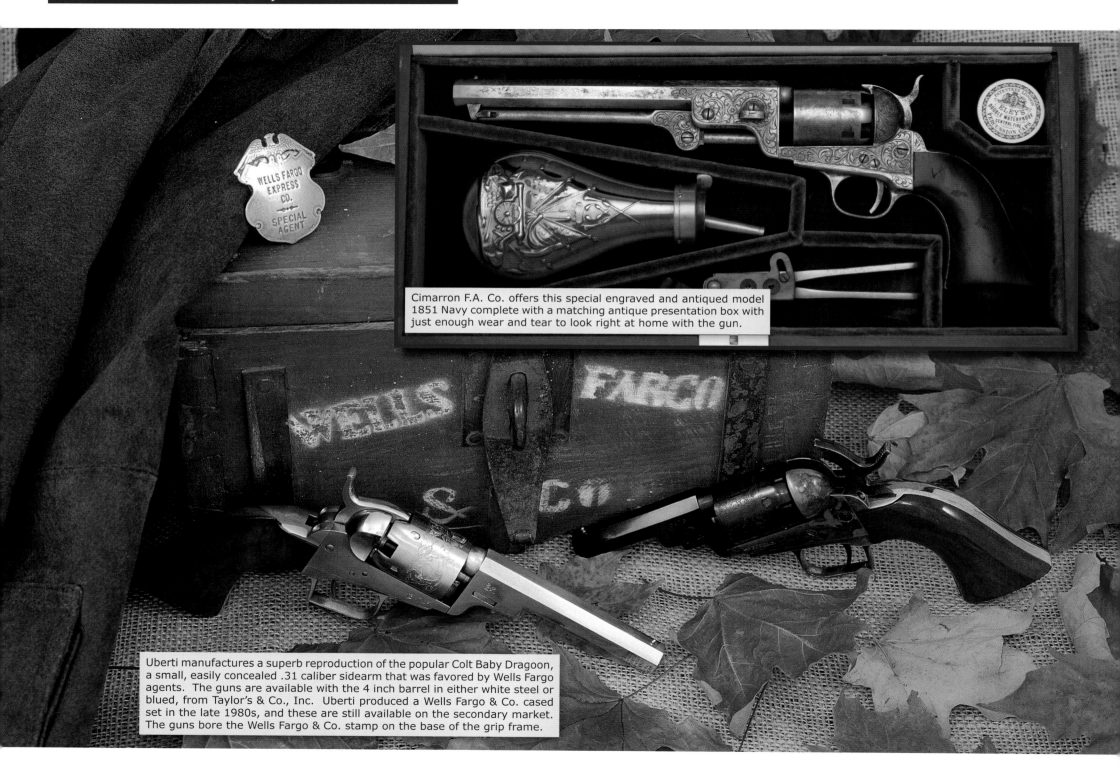

Cimarron F.A. Co. offers this special engraved and antiqued model 1851 Navy complete with a matching antique presentation box with just enough wear and tear to look right at home with the gun.

Uberti manufactures a superb reproduction of the popular Colt Baby Dragoon, a small, easily concealed .31 caliber sidearm that was favored by Wells Fargo agents. The guns are available with the 4 inch barrel in either white steel or blued, from Taylor's & Co., Inc. Uberti produced a Wells Fargo & Co. cased set in the late 1980s, and these are still available on the secondary market. The guns bore the Wells Fargo & Co. stamp on the base of the grip frame.

more aged appearance. Explains Suzanne Webb of A. Uberti, "The guns, in the white, are tossed into a tumbler with small rocks and bits of iron which take off the hard edges, giving the parts a worn appearance. The stocks are thrown in at the end to get a similar softening and wearing of the surface. The frame, barrel, and other exposed parts are then acid treated to give them a slightly corroded patina, dark blemishes and pits – the same kind of aging a gun might have after a century of wear and tear." The finish on the Uberti pistols varies from gun to gun with some looking worse than others, which in this instance, is a good thing.

Both the Pietta and Uberti revolvers look truly weathered, which seems to have great appeal among shooters desiring an authentic-looking but modern-made black powder pistol. These factory built old guns have led to many other prematurely aged copies of Civil War era guns since the late 1990s. It has almost become an art form in recent years.

Starting this trend certainly wasn't what Cimarron F.A. Co. president Mike Harvey had in mind when he set out to have the aged Colts produced by Uberti.

"I had come with the idea for an aged gun that could be sold to collectors who otherwise pay anywhere from $1,000 to $1,500 for an old, beat up 1851 Navy just to put it into a holster display," says Harvey. "They're for holster collectors who pay big bucks for original old gun leather and then want a gun to fill it out. I'd figured what would they do if they had a gun that looked old but only cost them $350 to $400 to stuff in a holster. I never expected it to catch on with the shooters," admits Harvey, still with a tone of disbelief in his voice. "They became the hottest item in our display at the 1998 Shot Show!" And a decade later they're still in demand. That pretty much shoots holes in everything we've said about fit and finish, at least if you're looking for a gun that appears to have been around since 1860.

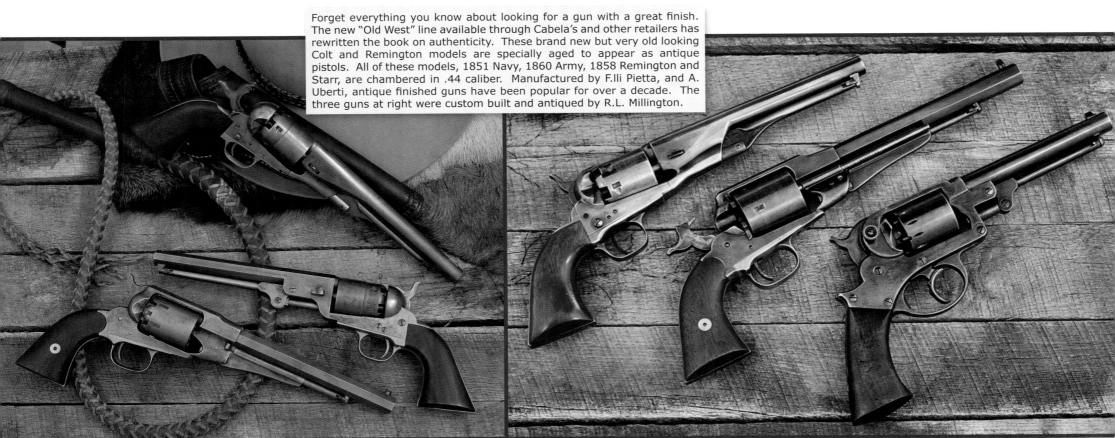

Forget everything you know about looking for a gun with a great finish. The new "Old West" line available through Cabela's and other retailers has rewritten the book on authenticity. These brand new but very old looking Colt and Remington models are specially aged to appear as antique pistols. All of these models, 1851 Navy, 1860 Army, 1858 Remington and Starr, are chambered in .44 caliber. Manufactured by F.lli Pietta, and A. Uberti, antique finished guns have been popular for over a decade. The three guns at right were custom built and antiqued by R.L. Millington.

CHAPTER THREE: Reproduction Percussion Pistols

Brass framed Navy models from F.lli Pietta include the Reb Confederate in either .36 or .44 caliber with half round, half octagonal barrel; Reb Nord Navy in .36 and .44 caliber with octagonal barrel; and the Navy Carbine with 12 inch octagonal barrel.

The Dragoon style worked well in every size, from the imposing Walker to the demure Pocket Model Navy of Caliber, and everything in between. Models pictured are manufactured by A. Uberti.

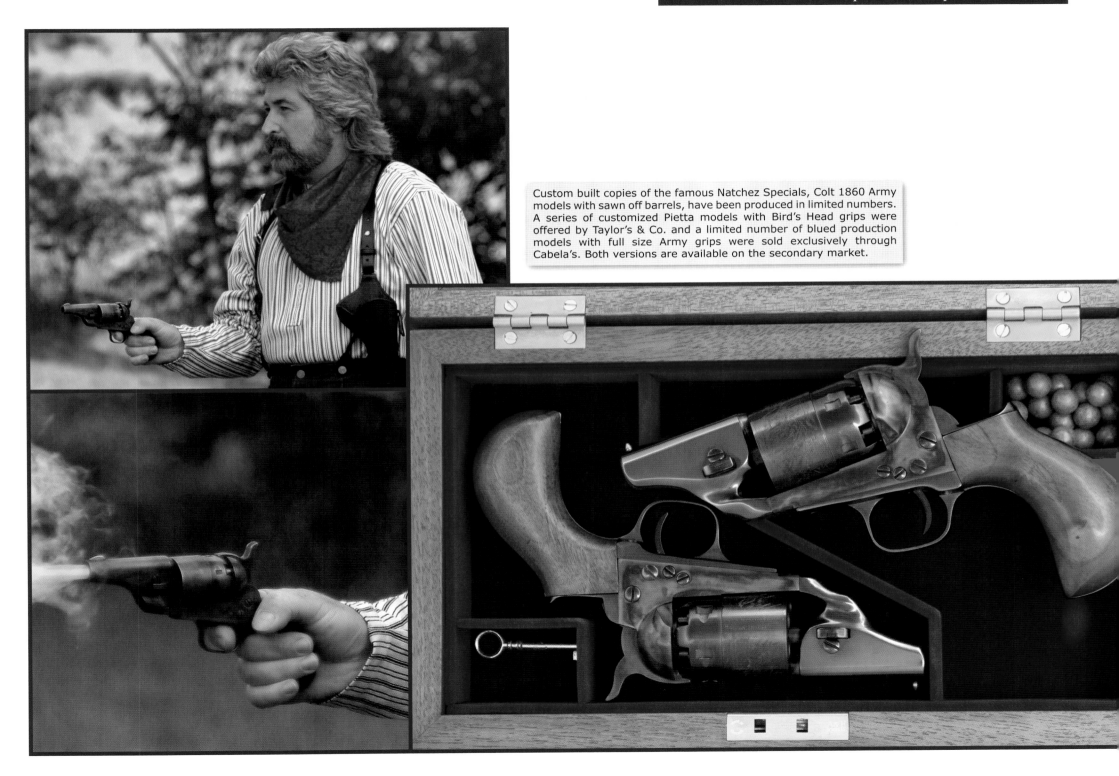

Custom built copies of the famous Natchez Specials, Colt 1860 Army models with sawn off barrels, have been produced in limited numbers. A series of customized Pietta models with Bird's Head grips were offered by Taylor's & Co. and a limited number of blued production models with full size Army grips were sold exclusively through Cabela's. Both versions are available on the secondary market.

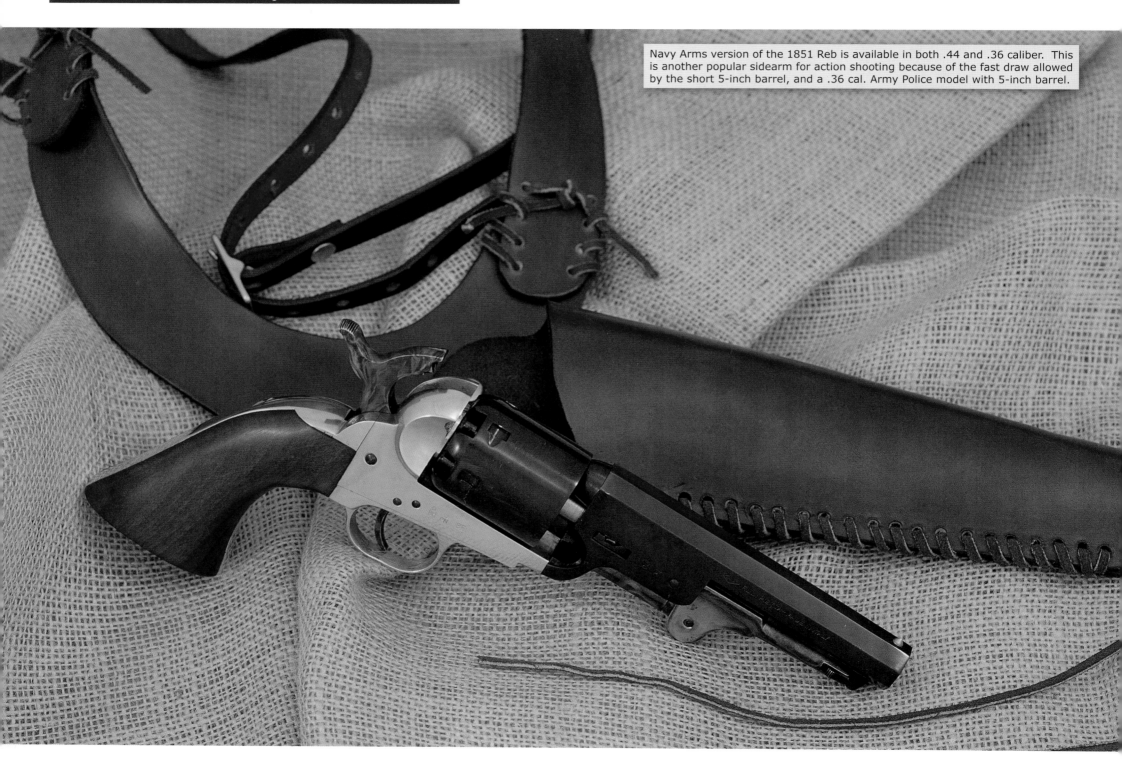

Navy Arms version of the 1851 Reb is available in both .44 and .36 caliber. This is another popular sidearm for action shooting because of the fast draw allowed by the short 5-inch barrel, and a .36 cal. Army Police model with 5-inch barrel.

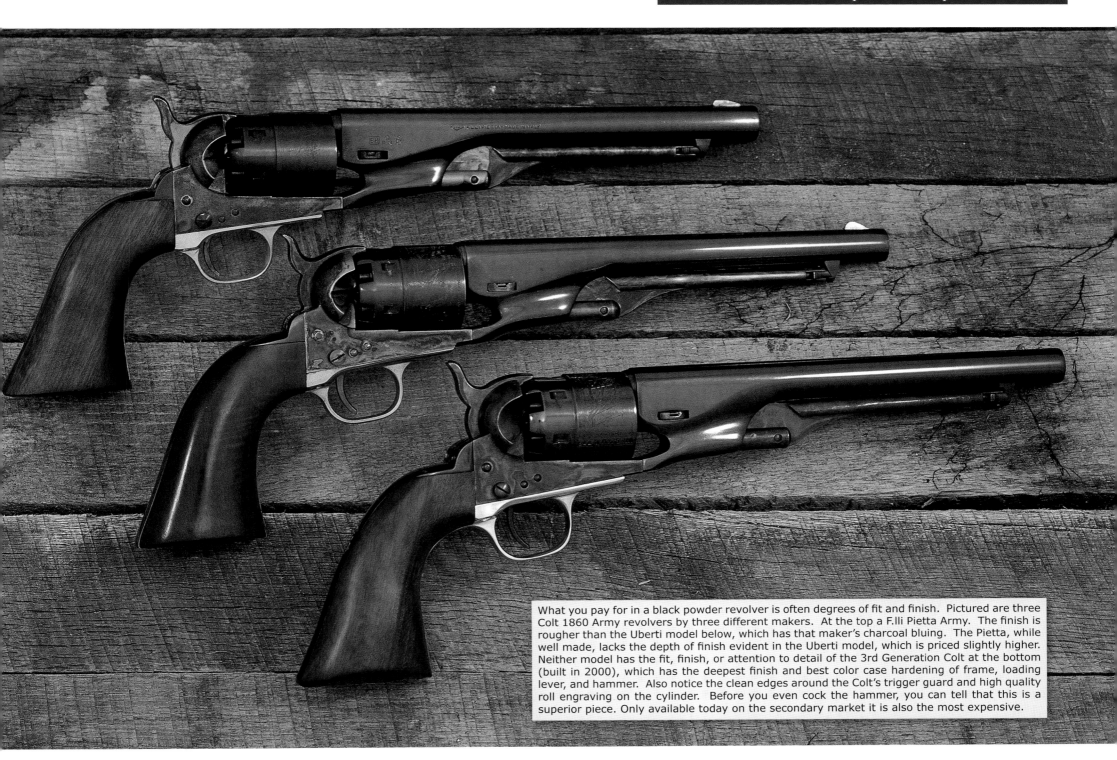

What you pay for in a black powder revolver is often degrees of fit and finish. Pictured are three Colt 1860 Army revolvers by three different makers. At the top a F.lli Pietta Army. The finish is rougher than the Uberti model below, which has that maker's charcoal bluing. The Pietta, while well made, lacks the depth of finish evident in the Uberti model, which is priced slightly higher. Neither model has the fit, finish, or attention to detail of the 3rd Generation Colt at the bottom (built in 2000), which has the deepest finish and best color case hardening of frame, loading lever, and hammer. Also notice the clean edges around the Colt's trigger guard and high quality roll engraving on the cylinder. Before you even cock the hammer, you can tell that this is a superior piece. Only available today on the secondary market it is also the most expensive.

CHAPTER FOUR: Limited Edition and Collectable Models

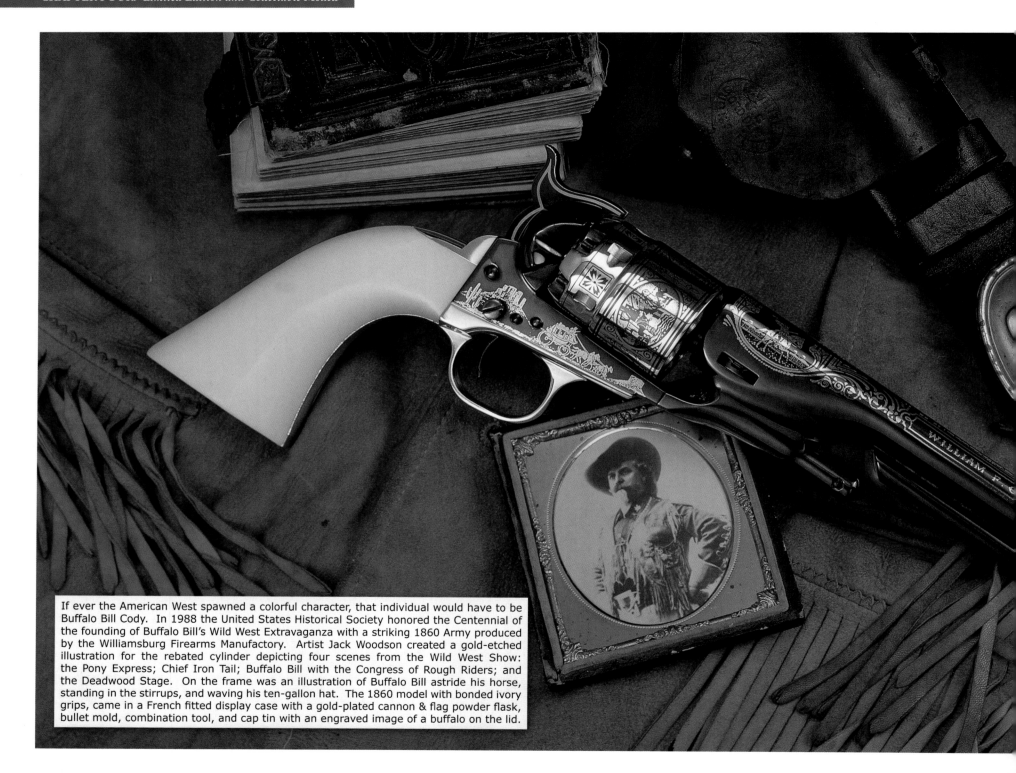

If ever the American West spawned a colorful character, that individual would have to be Buffalo Bill Cody. In 1988 the United States Historical Society honored the Centennial of the founding of Buffalo Bill's Wild West Extravaganza with a striking 1860 Army produced by the Williamsburg Firearms Manufactory. Artist Jack Woodson created a gold-etched illustration for the rebated cylinder depicting four scenes from the Wild West Show: the Pony Express; Chief Iron Tail; Buffalo Bill with the Congress of Rough Riders; and the Deadwood Stage. On the frame was an illustration of Buffalo Bill astride his horse, standing in the stirrups, and waving his ten-gallon hat. The 1860 model with bonded ivory grips, came in a French fitted display case with a gold-plated cannon & flag powder flask, bullet mold, combination tool, and cap tin with an engraved image of a buffalo on the lid.

LIMITED EDITION AND COLLECTABLE MODELS

Engraved and Boxed For Presentation
In The Colt Tradition

Throughout the 18th and 19th century, one of the finest gifts that could be bestowed upon a gentleman, or a lady, was a finely engraved presentation pistol. This tradition is as old as the firearm itself, and for generations before the advent of Samuel Colt's revolver, beautifully cased, engraved and embellished wheelock and flintlock pistols were exchanged as gifts among the aristocracy and heads of state.

These devices of otherwise fierce power and destruction were tamed and ennobled when touched by the hands of artisans practicing the time-honored craft of fine engraving. Arms and Armor authority Larry Wilson says that in collecting fine guns, "...the aficionado shares a passion that dates back to the original collectors, the British and continental European aristocrats who were not only shooters, but frequently devoted students and impassioned collectors. The collecting of valued objects, craftsmanship, and curiosity is an occupation as ancient as civilization itself."

Fortunately, today, one need not have a title in order to acquire such *objets d'art*. As decorative pieces, embellished firearms have joined fine furniture, silver, rare coins, and vintage automobiles as among the most highly valued objects in collecting. According to British art historian and author Graham Hood, "Pistols especially attained a rare harmony of graceful form and exquisite decoration in [the 17th century] and maintained it through the next. Rich patronage and fastidious skills combined to produce results that elicited pride and admiration then and now."

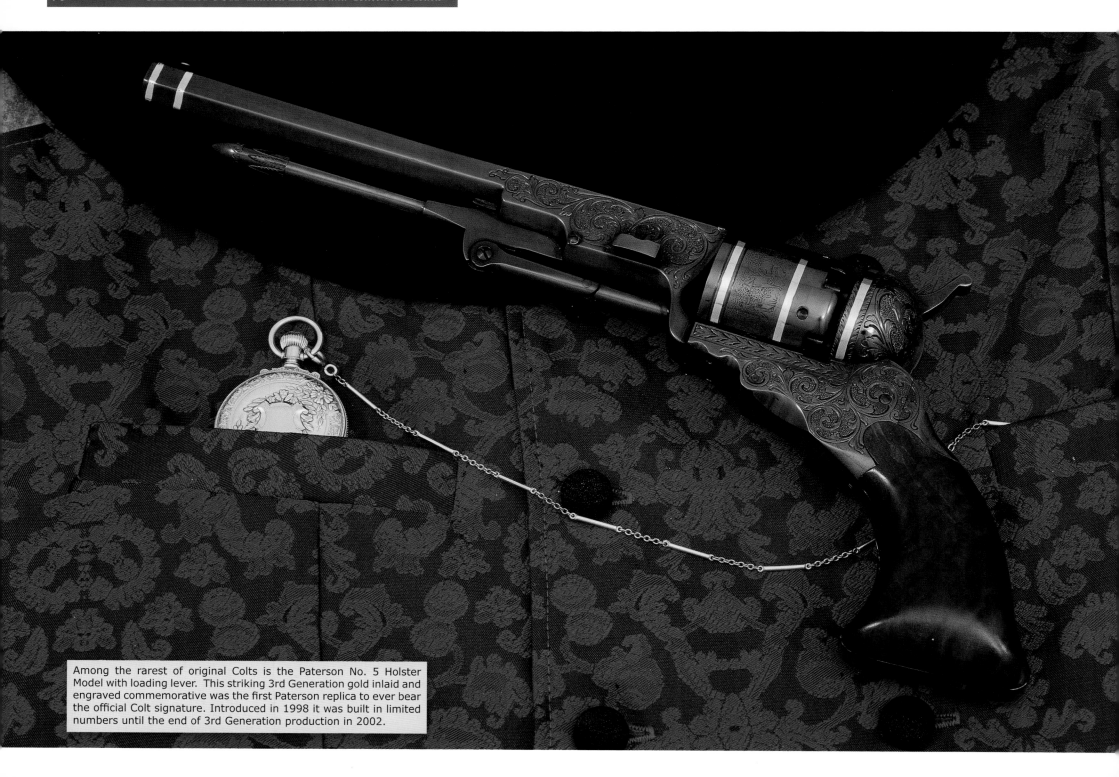

Among the rarest of original Colts is the Paterson No. 5 Holster Model with loading lever. This striking 3rd Generation gold inlaid and engraved commemorative was the first Paterson replica to ever bear the official Colt signature. Introduced in 1998 it was built in limited numbers until the end of 3rd Generation production in 2002.

Perhaps the amazing thing is that in over four hundred years the art of the gunsmith and engraver has not died out. The traditional symbol of the presentation piece has survived to call forth outstanding qualities of craftsmanship in modern day artisans. "Fine weapons, and the accessories connected with them," explains Hood, "are frequently beautiful and important, and deserve an integral place in any collection that attempts to reveal, through objects, the culture of an earlier age."

In the United States, firearms produced throughout the settling of the American West and Civil War era are the most coveted today. "During Samuel Colt's lifetime approximately 600,000 Colt percussion pistols of all types were manufactured and only a fraction of those were engraved presentation models," notes Wilson. "In his entire career, Colt may have given away as many as 3,500 presentation pistols to military officers, foreign dignitaries, and heads of state." These 19th-century arms have become the most prized Colts of the 20th and 21st centuries, with many cased, engraved models valued today at well over $1 million. The earliest Colt to be engraved was produced late in the 1830s. It was a Paterson, Sam Colt's first revolver. The best of these, an historic five-shot, .36 caliber cap & ball pistol, inspired Colt Blackpowder Arms to produce a limited contemporary version of the now legendary No. 5 Holster Model or "Texas Paterson," a name bestowed upon the gun for its heralded use by the Texas Rangers in the early 1840s. Not more than about 100 original Holster Models with the attached loading lever were produced, and of those only a small number were factory engraved presentation pistols. The new No. 5 Holster Model became the first authentic Colt Paterson revolver to be made in more than 150 years.

The Colt Blackpowder Arms model faithfully rendered the design of an early cased presentation piece produced in 1842 with inlaid silver bands (done here in 18-karet gold) accenting a beautifully engraved barrel, cylinder, and recoil shield. The reproduction Colt Patersons bore the same hand-engraved vine scroll as the original, with punch-dot background on the barrel, frame, backstrap, recoil shield, and three-piece loading lever. It was accurate in every detail down to the original "Patent Arms Mg.Co. Paterson, NJ-Colt's PT" barrel address and Stagecoach Holdup roll engraved cylinder scene. Each bore authentic Colt serial numbers on all appropriate parts making it as close to the original as one can get without spending several hundred thousand dollars or more. That makes an authentic Colt reproduction seem like a bargain at under $5,000.

Perhaps the most remarkable of all Colt presentations were those embellished with silver and gold, and fitted with ornate silver- or gold-plated cast bronze Tiffany-

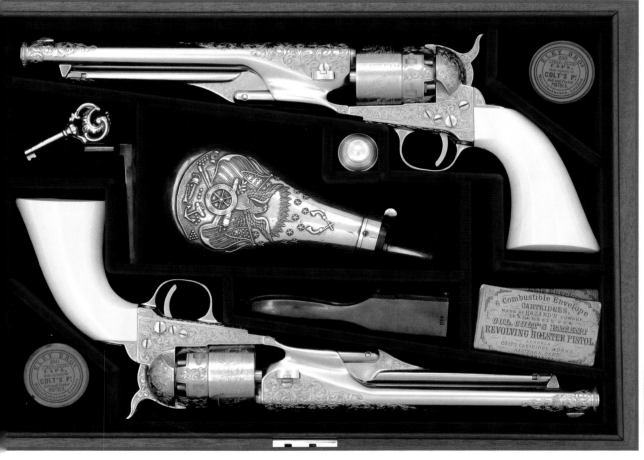

Copied from an original pair of cased and engraved Colt 1860 Army revolvers, John J. Adams, Sr. engraved this set as a private commission. The mahogany case was handcrafted from the original design by Pennsylvania furniture maker Duncan Everhart. The special accessories were done by Frank Klay.

CHAPTER FOUR: Limited Edition and Collectable Models

In 1971, Colt issued a pair of Civil War commemoratives, the Ulysses S. Grant and Robert E. Lee "Blue and Grey" set, two cased revolvers with period accessories. Serial numbers 582 (Lee) and 2141 (Grant). The pair was also sold in an even more exclusive double gun casing. (cased pair from Dennis Russell collection)

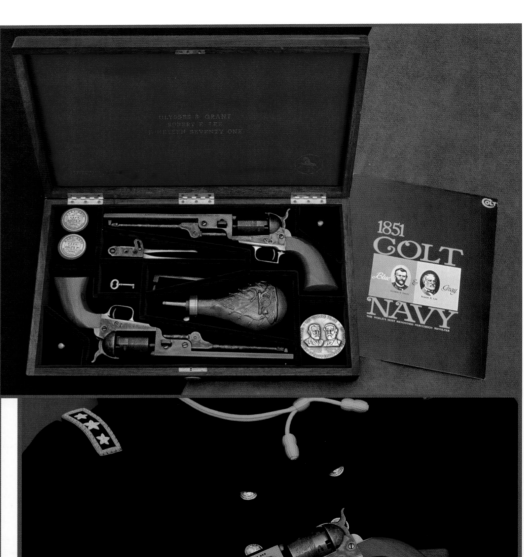

style stocks. One of the original engraved 1860 Tiffany-gripped revolvers, upon which the Colt Heirloom Edition Tiffany reproduction was based, would sell in the high six figures today in mint condition. This contemporary Colt was still a hefty $5,000 and required four months to make. It was manufactured in the same fashion, and using the same painstaking hand engraving techniques, as the original Tiffany-gripped revolvers produced for Colt 148 years ago. Even these reproductions are a rare find today.

"It surprises many people to find out that Tiffany is one of the oldest and most respected names in firearms embellishment in the world," says Wilson. "Tiffany designed arms have been a tradition with the renowned jeweler and silversmith since the 1860s."

William R. Chaney, Chairman, Tiffany & Co., noted that "Since 1837, Tiffany & Co. has been recognized around the world as a premier source of fine jewelry and silver creations. However, the range of the company's design expertise also extends to the realm of decorative arms.

"During the Civil War, many savvy retailers such as Charles Lewis Tiffany began supplying military equipage to the Union Army. Although the company had established itself as a premier source for household silver by this time, the war had adversely affected sales of such items. Making the best of the situation, Tiffany began applying its silversmithing techniques to presentation swords and firearms.

"The elaborate designs and the variety of techniques used in Tiffany swords made the firm America's preeminent swordmaker. Over the next fifty years, Tiffany's decorative firearms would achieve the same artistic status as its swords."

A new model introduced in 2007 by America Remembers was the Buffalo Bill Cody Wild West Show commemorative celebrating the 90th anniversary of the last Wild West Show staged in 1917. The gun, designed by the author, Paul Warden, and the craftsmen of A&A engraving features beautiful gold etched images of Cody and a Wild West Show banner along the barrel.

Legendary firearms engraver A.A. White and his associate Andrew Bourbon produced this stunning 3rd Model Colt Dragoon on a 2nd Generation Colt for the author in 1997. The gun features Gustav Young vine scroll coverage and ivory grips with a silver engraved cartouche on the left stock, and an engraved American eagle on the right, signed by A.A. White.

In his book *Steel Canvas*, which chronicles the history of engraved firearms, Wilson notes that Tiffany produced a significant number of engraved Colt pistols. The New York firm continued to produce presentation pistols from the mid-1880s to the early 1900s, and catalogs for 1900 through 1909 advertised *Revolvers of the most improved types, mounted in silver, carved ivory, gold, etc., with rich and elaborate decorations.* Wilson, who was also president of American Master Engravers, Inc., notes that Tiffany & Co. still produced a limited number of engraved and embellished pistols each year well into the late 20th century. Tiffany briefly revived their exclusive art of engraving fine pistols in the early 1980s, much through the efforts of George A. Strichman, then chairman of Colt Industries, who commissioned three gold and silver mounted Tiffany Colts. "That really marked the revival of the Tiffany tradition," says Wilson.

Although modern-day firearms are offered with gold inlay and engraving and feature gold and silver Tiffany stocks, many of the firm's most extraordinary examples have been based on the Colt 2nd Generation reproductions of Civil War era percussion pistols, such as the one commissioned in 1988 for Gene Autry's 81st birthday, a stunning Third Model Dragoon embellished in gold and silver overlay and inlay, with simulated snake skin vermeil stocks.

Another equally striking artistic achievement was the Tiffany & Co. Little Big Horn Dragoon designed for Larry Wilson. The engravings and illustrations on this piece were based on American Indian pictograph paintings of the battle as recorded by Red Horse and Kicking Bear, of the Sioux Nation, and Two Moon, of the Cheyenne. This remarkable pistol depicted Little Big Horn from the Indian's spiritual perspective.

"One of the most important developments in 20th century gun engraving occurred in 1971, when Colt returned black powder pistols to their product line through an arrangement with, among others, noted gunmakers Lou and Anthony Imperato, in New York. As a result, some of the finest antique Colts ever made have been produced in last 40 years," says Wilson. "There is a great interest in embellished guns today. This is really like

CHAPTER FOUR: Limited Edition and Collectable Models

The Colt Heritage 1847 Walker Dragoon was one of the finest commemorative pistols ever produced by Colt. The limited, cased model came with a special leather-bound and gilt edged edition of R.L. Wilson's book *The Colt Heritage*. A total of 1,853 were produced through June 1981. Serial number 166.

One of the most ornate percussion pistols to come from Colt was the Cochise, a mammoth Third Model Dragoon embellished with 24-kt. gold inlay and 24-kt. gold-plated cylinder.

the golden age of gun making when it comes to decoration, and by Colt Blackpowder Arms recreating and reissuing some of these earlier rarities until production ended in 2002, presented something for the collector and the shooter at a price range he could afford, whereas the original antiques have continued to hit record prices in many categories and therefore become extremely difficult to purchase without paying a substantial price."

While the majority of Colt reproductions are of the commercial style – those sold to the general public in the late 19th century – and used today for movies, Civil War reenactments, and Frontiersman competition in Single Action Shooting Society events, the modern-day Colt presentation pieces are nearly as rare as originals. A few have been produced through the Colt Custom Shop over the years, a few even by Tiffany & Co. Hand-crafted firearms were also commissioned through American Master Engravers in Hadlyme, Connecticut – the renowned firm headed by the late A.A. White – which for many years assisted Tiffany & Co. with highly exclusive presentation pistols.

Others have been masterfully created over the years by America Remembers (formerly the United States Historical Society), The American Historical Foundation, and by master engravers such as John J. Adams, Sr. and John Adams, Jr. of Adams & Adams in Vershire, Vermont.

Knowing the history behind the guns, the percussion revolvers that were manufactured by Colt Blackpowder seem more akin to continuations of original guns than reproductions. "Colt's limited editions, the Paterson and Heirloom in particular, definitely have future collectability," believes Wilson. "When you have a very limited number being produced, less than 100, they are collectible as soon as they are made." This has been the case with Colt Blackpowder pistols for more than 35 years. Many of the earlier models produced by Colt in the 1970s and early 1980s, such as the Heritage 1847 Walker, have become modern-day collectibles at an affordable price.

Engraving of good quality is reasonably priced, and the firearm when

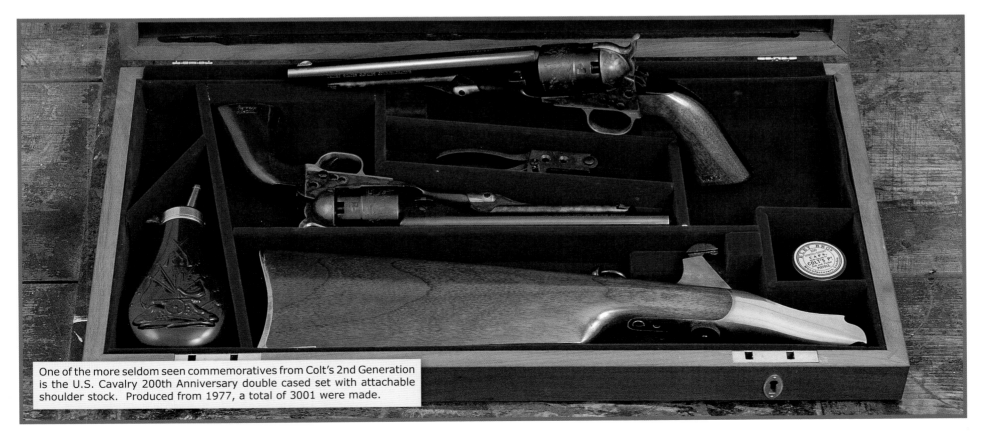

One of the more seldom seen commemoratives from Colt's 2nd Generation is the U.S. Cavalry 200th Anniversary double cased set with attachable shoulder stock. Produced from 1977, a total of 3001 were made.

CHAPTER FOUR: Limited Edition and Collectable Models

completed is often worth more than the combined cost of the gun and its decoration. The skilled artisans in Gardone and Brescia, Italy, are perhaps the finest engravers in the world. This has allowed companies like A. Uberti and F.lli Pietta, to offer superbly hand-crafted, engraved black powder pistols for more than two decades. Pietta currently has more engraved models available than any percussion gunmaker in Italy, including a striking white metal LeMat, and variations of the 1860 Army, 1851 Navy, and 1858 Remington, among others.

Contemporary artisans like John J. Adams, Sr. still use the same fundamental hand tools and venerable skills as 19th century engravers like Gustave Young, Louis Nimschke, and Cuno A. Helfricht, all of whom created some of the most outstanding presentation engraved firearms in the world. The rich, strongly Germanic style of these engravers still dominates American arms engraving today and is popularly called "the American style." Original firearms by Young and Nimschke command six figures. "There is a realization among collectors," observes Wilson, "that quality decorative firearms, old and new, represent good investments."

Another 2nd Generation cased model was the 1862 Pocket Model of Navy Caliber, limited to a series of 500 guns in a French fitted black walnut box with pewter flask, bullet mold, combination tool, and cap tin. Produced from December 1979 to November 1981. Serial number 50611.

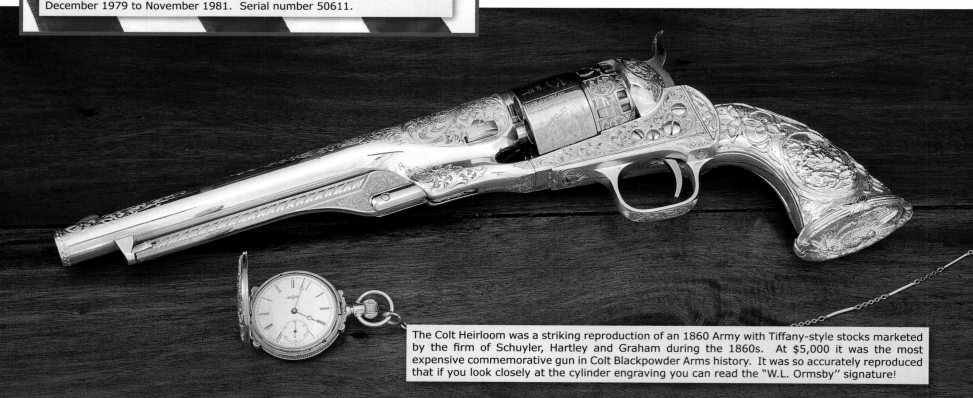

The Colt Heirloom was a striking reproduction of an 1860 Army with Tiffany-style stocks marketed by the firm of Schuyler, Hartley and Graham during the 1860s. At $5,000 it was the most expensive commemorative gun in Colt Blackpowder Arms history. It was so accurately reproduced that if you look closely at the cylinder engraving you can read the "W.L. Ormsby" signature!

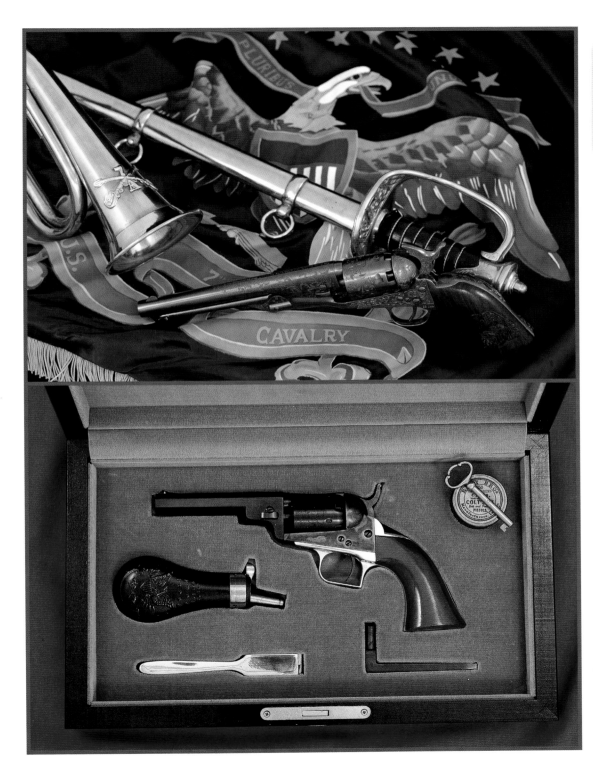

Chambered in .36 caliber, the George Armstrong Custer 3rd Generation 1861 Navy had an antique silver blue finish embellished in the style of Louis Nimschke, one of the premiere engravers of Colt revolvers in the second half of the 19th century. Presentation stocks for the Custer were select rosewood, engraved with a carved American eagle over a shield on the left-hand side, with a deep checkered pattern appropriate to the era on the right. Reproduction of Custer's 7th Cavalry flag courtesy Hugh Tracy collection.

White, who continued to engrave almost up until the time of his death was a modern day Gustave Young. "In the annals of firearms engravers, he was one of the greatest in history," says Wilson. "He probably did more guns for famous people than any other American engraver in the 20th century." White created engraved pistols for Tiffany & Co., heads of state, and U.S. Presidents John F. Kennedy, Lyndon B. Johnson, Gerald Ford, and Ronald Reagan.

Over the decades since Colt's began remanufacturing models from the 19th century and continuing its production of the 1873 Peacemaker, thousands of extraordinary guns have been produced. Many of the finest ever done in the mid to late 20th century were by White, his protégé Andrew Bourbon, Winston Churchill, Howard Dove, Ken Hurst, George Spring, Denise Thirion, Leonard Francolini, and K. C. Hunt.

To watch master engravers at work, like John J. Adams, Sr., who was one of the principal engravers for A. A. White during the Strichman era, is an experience. Andrew Bourbon continues to engrave in the various classic styles, following design patterns created more than a century ago by Young and Nimschke for Colt and Smith & Wesson, but engraves like Bourbon an Adams also create their own styles, which in almost every instance rival or exceed the greatest works of master engravers dating back five centuries.

The last model introduced by Colt Blackpowder in the 2nd Generation was the 1847-48 Baby Dragoon, a pocket-sized version of the 1st Model Dragoon chambered in .31 caliber. The gun was first introduced in 1979 as a limited edition of 500 presentation models each in a French fitted case with powder flask, bullet mold, percussion cap tin and combination tool. Serial number 16501.

CHAPTER FOUR: Limited Edition and Collectable Models

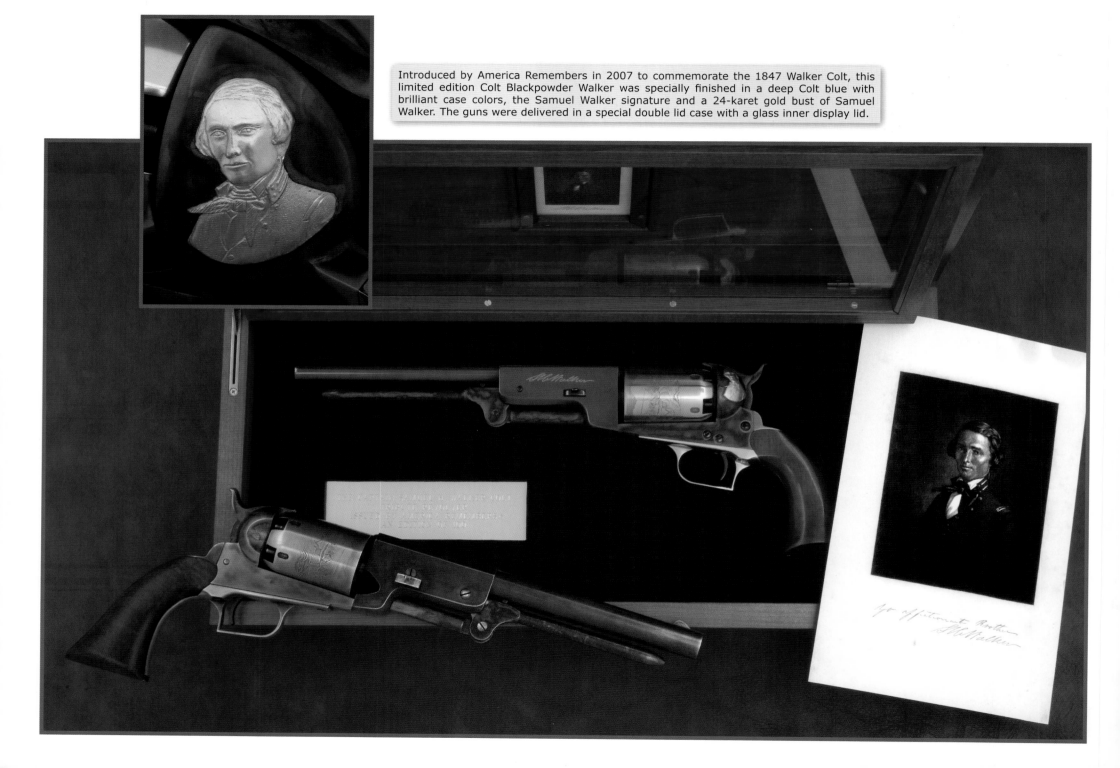

Introduced by America Remembers in 2007 to commemorate the 1847 Walker Colt, this limited edition Colt Blackpowder Walker was specially finished in a deep Colt blue with brilliant case colors, the Samuel Walker signature and a 24-karet gold bust of Samuel Walker. The guns were delivered in a special double lid case with a glass inner display lid.

Introduced in 1997 to commemorate the sesquicentennial of the 1847 Walker Colt, the 3rd Generation "150th Anniversary Walker Dragoon," was issued with gold "A Company No. 1" markings on the barrel lug, frame and cylinder to commemorate the Walker's use by the Texas Rangers and U.S. Mounted Rifles in the war with Mexico. The original A Company models ended with serial number 220 and the 150th Anniversary Walkers began with number 221. The gun pictured is the first produced in the series, and has been affixed with a special gold engraved cartouche.

In a way it is harder for engravers today because of changes in metallurgy, use of investment castings, and the hardness and toughness of steel. In the days of the Old West the steel was dead soft and it was relatively easy to cut with an engraver's tool. "In order to do engraving today," says Wilson, "you need to have real expertise in metallurgy to soften and then reharden metal, otherwise you can compromise the integrity of the firearm, and fundamentally, all of these engraved guns are supposed to be functional first."

The variety of engraved black powder pistols available today truly rivals those of Samuel Colt's own era. There is an air of exclusivity to having the finest of anything, and that includes firearms, old or new. Wilson finds it fascinating that the majority of collectors today like modern firearms – those produced in the 20th century. "I think engraving is the primary reason. The work today is spectacular, not only the engraving, gold and silver inlay and sculpting, but the quality of the firearms themselves." Wilson notes that revolvers are sometimes supplied with a wooden dowel down the barrel and into the cylinder to prevent the mechanism from being worked. Decorative arms, even when new, were seldom discharged by their owners. Even in an era when the carrying of sidearms was *de rigueur*, embellished presentation pieces were regarded as art objects to be admired and displayed.

In Sam Colt's lifetime there were only about 600,000 Colt firearms made and another 400,000 percussion types after his death, many of which were destroyed during the Civil War or have deteriorated over the years. There are a finite number of exceptional examples left in the world today that were embellished by famous engravers. Ironically, the names of 19th century engravers were barely known by collectors until Larry Wilson's first book, *Samuel Colt Presents*, was published in 1961.

"When I began researching the book I started bumping into people who were descendants from these great engravers. In time I found Gustave Young's grandson, Helfricht's

CHAPTER FOUR: Limited Edition and Collectable Models

Master engraver John J. Adams, Sr. recreates a Nimschke style motif on a Colt 1860 Army. Adams still uses the same fundamental hand tools and venerable skills as 19th century engravers like Gustave Young, Louis Nimschke, and Cuno A. Helfricht.

granddaughter, and others. Fortunately, the Germans never throw anything away, and Young's grandson had an attic full of documents, memorabilia and even rare original design patterns." Wilson also authored a book on Louis Nimschke's designs in 1964, and single-handedly attached names to engraving styles giving countless collectable firearms provenance, creating a reference for many of the engraved designs seen on today's reproduction Colt black powder revolvers. Among the most sought after originals today are presentation Colts from the 1850s and 1860s. In an article published in the November 1996 issue of *Art & Antiques* magazine, renowned collector and auction house owner Greg Martin noted that, "A 19th century Colt will bring far more than any highly decorated 17th or 18th century European firearm."

As collectibles, Martin says that guns actually outpaced many other categories handily in the 1990s. "The gold inlaid guns that are authentic of the period and that have a verifiable history are the absolute ultimate for collectors," adds Wilson. "There are still a fair number of Colts from the percussion period that are available, although it's pretty rarified at the top." This is one reason that engraved Colt reproductions have remained popular among collectors.

This increasing demand for quality workmanship on new pistols, both here and in Europe also ensures that the celebrated art of the firearms engraver and the legendary Colt models of the 19th century shall not perish with the passing of time.

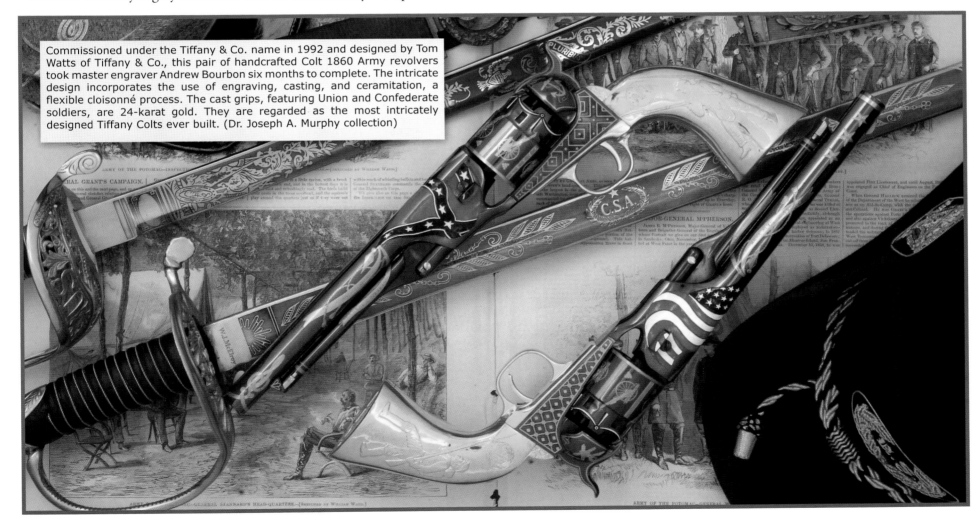

Commissioned under the Tiffany & Co. name in 1992 and designed by Tom Watts of Tiffany & Co., this pair of handcrafted Colt 1860 Army revolvers took master engraver Andrew Bourbon six months to complete. The intricate design incorporates the use of engraving, casting, and ceramitation, a flexible cloisonné process. The cast grips, featuring Union and Confederate soldiers, are 24-karat gold. They are regarded as the most intricately designed Tiffany Colts ever built. (Dr. Joseph A. Murphy collection)

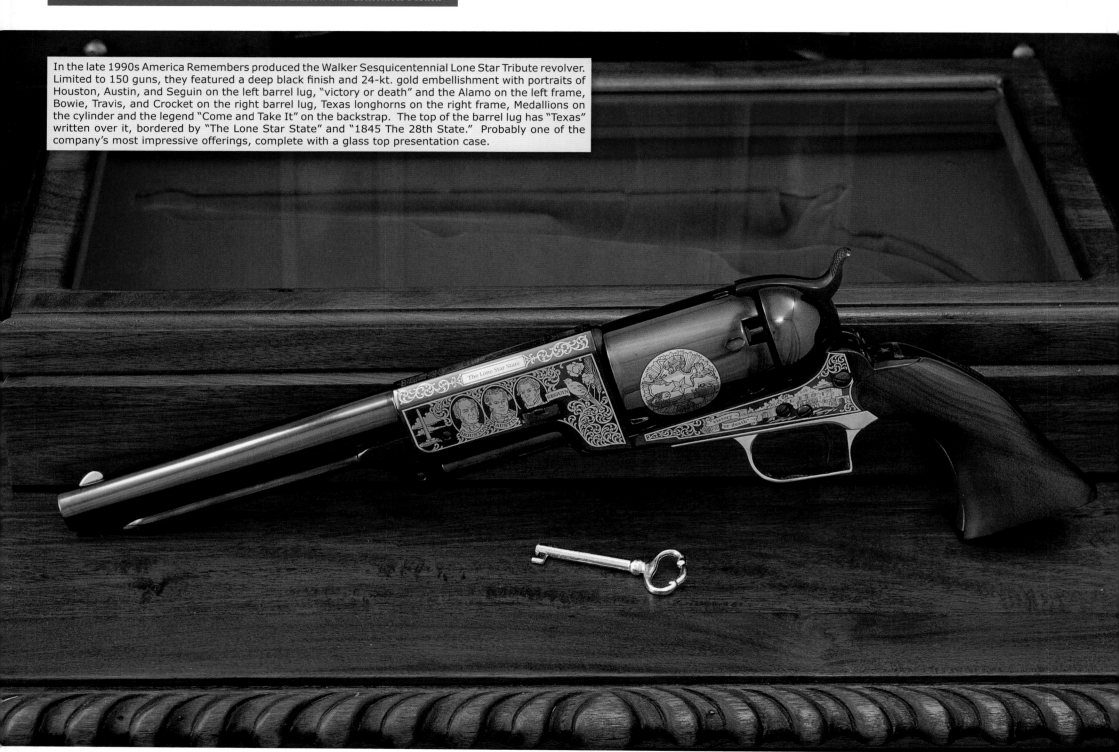

In the late 1990s America Remembers produced the Walker Sesquicentennial Lone Star Tribute revolver. Limited to 150 guns, they featured a deep black finish and 24-kt. gold embellishment with portraits of Houston, Austin, and Seguin on the left barrel lug, "victory or death" and the Alamo on the left frame, Bowie, Travis, and Crocket on the right barrel lug, Texas longhorns on the right frame, Medallions on the cylinder and the legend "Come and Take It" on the backstrap. The top of the barrel lug has "Texas" written over it, bordered by "The Lone Star State" and "1845 The 28th State." Probably one of the company's most impressive offerings, complete with a glass top presentation case.

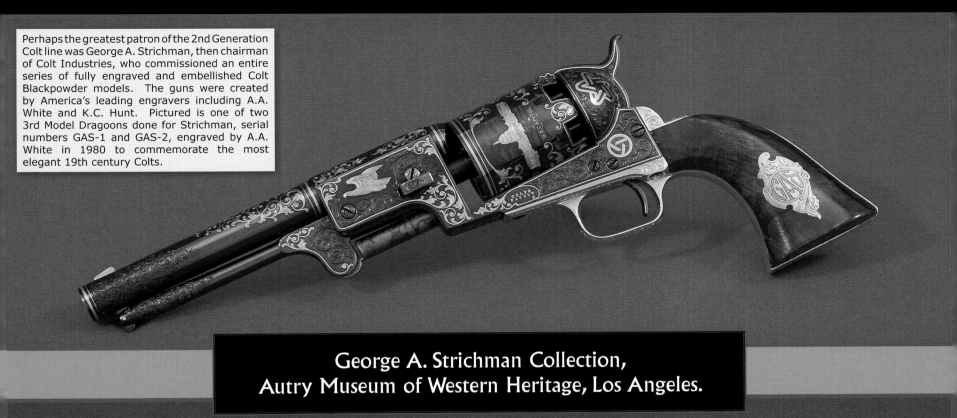

Perhaps the greatest patron of the 2nd Generation Colt line was George A. Strichman, then chairman of Colt Industries, who commissioned an entire series of fully engraved and embellished Colt Blackpowder models. The guns were created by America's leading engravers including A.A. White and K.C. Hunt. Pictured is one of two 3rd Model Dragoons done for Strichman, serial numbers GAS-1 and GAS-2, engraved by A.A. White in 1980 to commemorate the most elegant 19th century Colts.

George A. Strichman Collection, Autry Museum of Western Heritage, Los Angeles.

Silver and gold 3rd Model Dragoon was engraved by A.A. White. A matched pair were produced for Strichman in Gustav Young-style scrollwork.

CHAPTER FOUR: Limited Edition and Collectable Models

The American Historical Foundation produced 500 J.E.B. Stuart commemorative LeMat Cavalry revolvers. Black finish with 24-kt. gold plating. The black finish was symbolic of mourning, a reminder that General Stuart died in the war between the states. This was the first Limited Edition LeMat ever produced.

Another superb 1860 model from the United States Historical Society was the U.S. Cavalry Pistol, embellished with two Jack Woodson scenes around the rebated cylinder, United States Cavalry and crossed sabers along the length of the barrel, and magnificent genuine stag grips. This edition was limited to 975 examples.

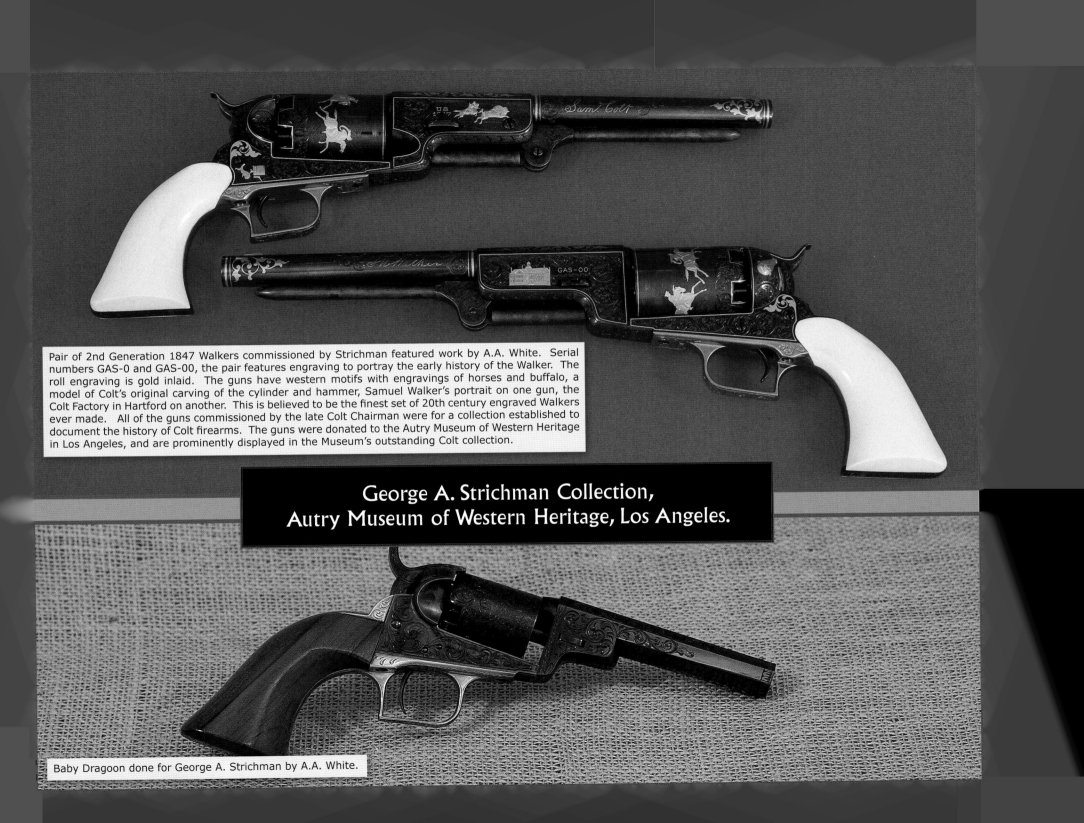

CHAPTER FOUR: Limited Edition and Collectable Models

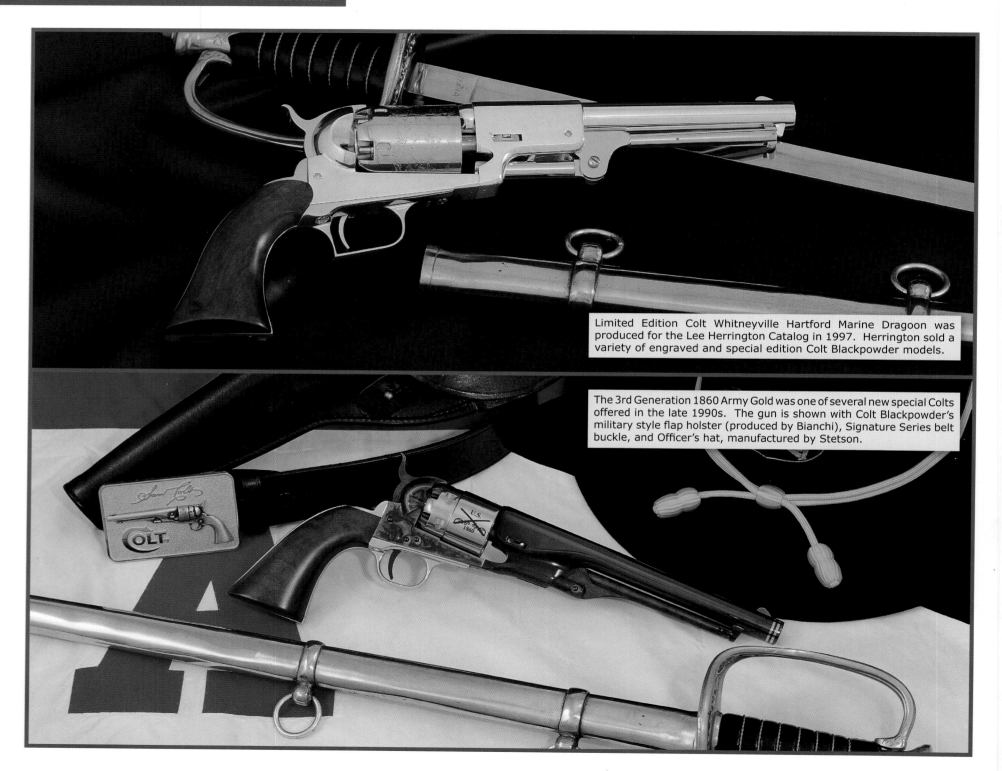

Limited Edition Colt Whitneyville Hartford Marine Dragoon was produced for the Lee Herrington Catalog in 1997. Herrington sold a variety of engraved and special edition Colt Blackpowder models.

The 3rd Generation 1860 Army Gold was one of several new special Colts offered in the late 1990s. The gun is shown with Colt Blackpowder's military style flap holster (produced by Bianchi), Signature Series belt buckle, and Officer's hat, manufactured by Stetson.

California Commemorative Dragoon, serial number 25001 was manufactured in 1982 and factory engraved by Howard Dove. The gun was commissioned by holster-maker John Bianchi to portray California history and had 100 percent coverage. The cylinder bears the Great Seal of the State of California.

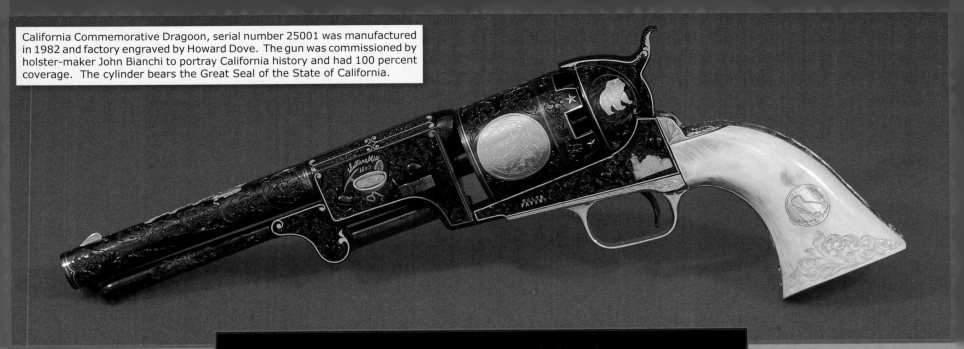

George A. Strichman Collection, Autry Museum of Western Heritage, Los Angeles.

America Remembers introduced the Heritage Series Paterson in 2002. Designed by the author and R. L. Millington, the first series Heritage was offered in antique blue (shown) charcoal blue and antique gray in a walnut presentation case with accessories. It was followed by the deluxe engraved, silver inlaid and ivory gripped Improved Belt Model No. 2 in 2004. The model with flared grips is a one-off-version done by Dan Chesiak (who produced the ivory grips for America Remembers) and features special engraving and silver inlay work.

CHAPTER FOUR: Limited Edition and Collectable Models

Before the Civil War, Jefferson Davis was U.S. Secretary of War and in 1858 Samuel Colt presented him with a handsomely engraved, shoulder stocked 1851 Navy revolver. This example is one of 250 limited edition reproductions produced by The U.S. Historical Society in 1990.

During the manufacture of 2nd Generation Colt Blackpowder models, a special series of 2,945 cased pairs of 1860 Army revolvers, with a single shoulder stock and accessories, were produced as the "U.S. Cavalry Commemorative" model. Within this series, between 1977 and 1980, 40 pair were to be custom engraved and gold inlaid. Although this example is marked No. 11 of 40, only 23 pair were ever completed. This has become one of the more rare of 2nd Generation cased sets.

CHAPTER FOUR: Limited Edition and Collectable Models

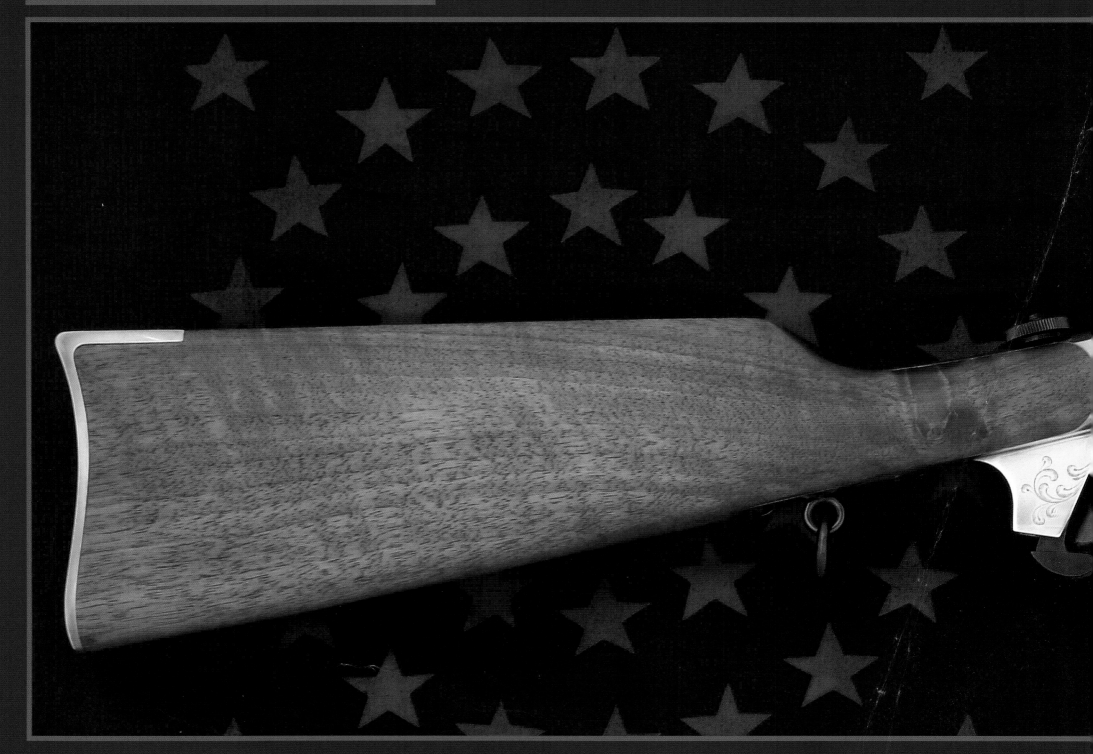

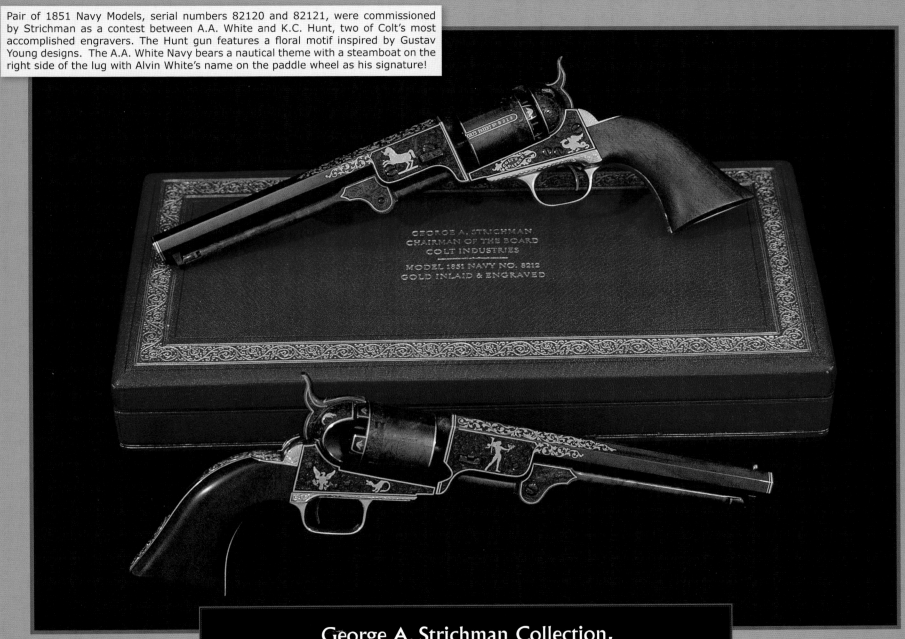

Pair of 1851 Navy Models, serial numbers 82120 and 82121, were commissioned by Strichman as a contest between A.A. White and K.C. Hunt, two of Colt's most accomplished engravers. The Hunt gun features a floral motif inspired by Gustav Young designs. The A.A. White Navy bears a nautical theme with a steamboat on the right side of the lug with Alvin White's name on the paddle wheel as his signature!

George A. Strichman Collection,
Autry Museum of Western Heritage, Los Angeles.

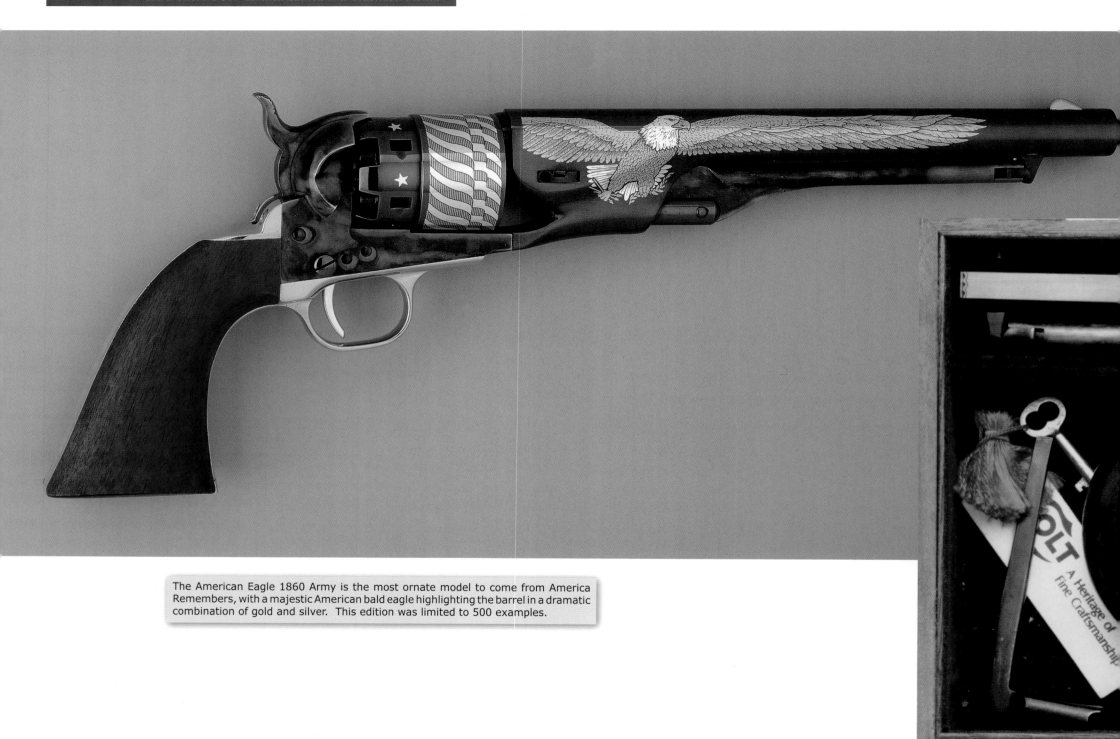

The American Eagle 1860 Army is the most ornate model to come from America Remembers, with a majestic American bald eagle highlighting the barrel in a dramatic combination of gold and silver. This edition was limited to 500 examples.

Paramount among specially engraved Limited Editions were the 1851 Klay Navy models engraved by Daniel Cullitty in the Gustav Young Style. Notes R.L. Wilson of the models produced by Frank Klay, "Klay Colts have proven popular with antique arms collectors, as well as devotees of modern gunmaking, their quality standards being strictly equal to those of products from Sam Colt's own lifetime."

CHAPTER FOUR: Limited Edition and Collectable Models

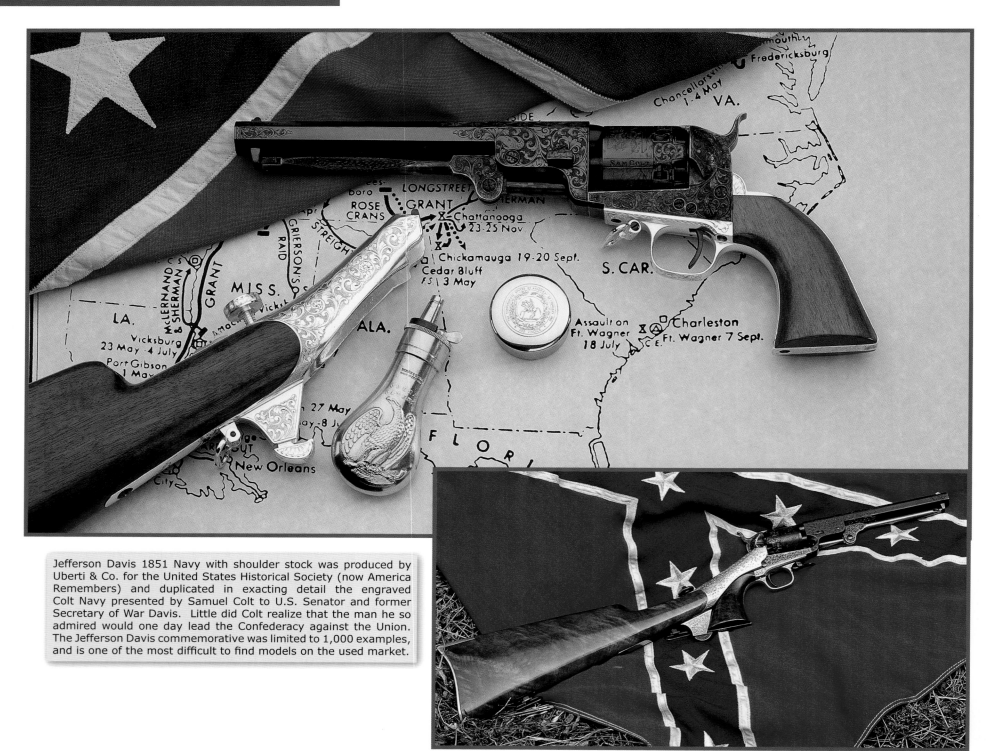

Jefferson Davis 1851 Navy with shoulder stock was produced by Uberti & Co. for the United States Historical Society (now America Remembers) and duplicated in exacting detail the engraved Colt Navy presented by Samuel Colt to U.S. Senator and former Secretary of War Davis. Little did Colt realize that the man he so admired would one day lead the Confederacy against the Union. The Jefferson Davis commemorative was limited to 1,000 examples, and is one of the most difficult to find models on the used market.

Black Powder Revolvers - Reproductions & Replicas

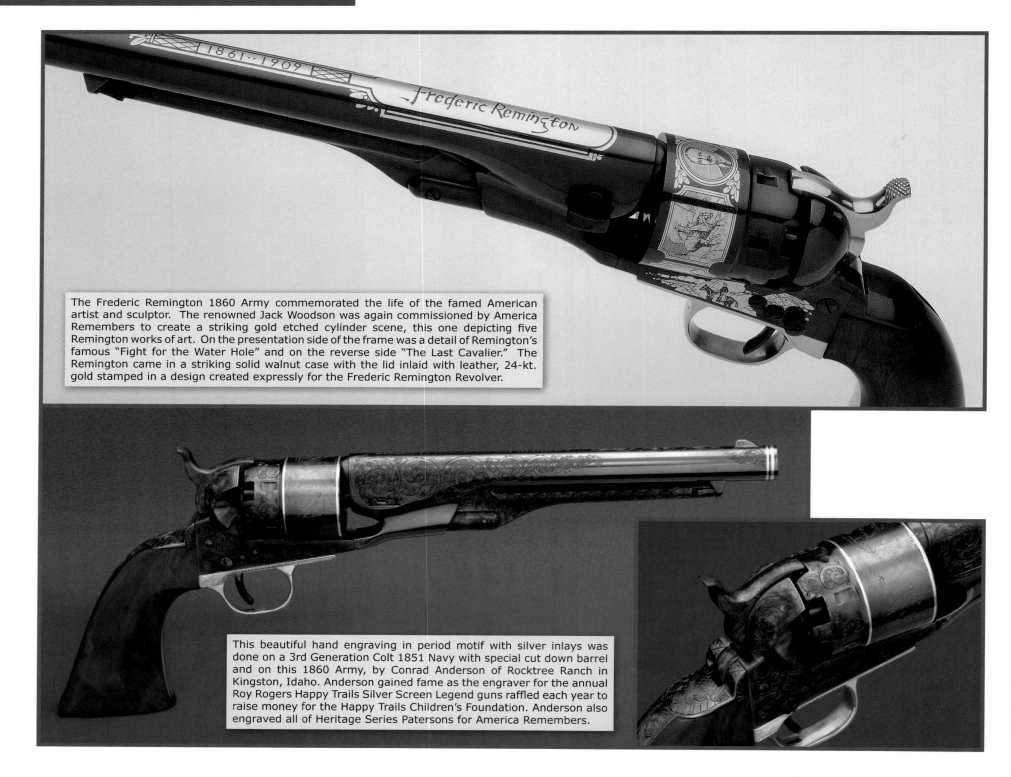

The Frederic Remington 1860 Army commemorated the life of the famed American artist and sculptor. The renowned Jack Woodson was again commissioned by America Remembers to create a striking gold etched cylinder scene, this one depicting five Remington works of art. On the presentation side of the frame was a detail of Remington's famous "Fight for the Water Hole" and on the reverse side "The Last Cavalier." The Remington came in a striking solid walnut case with the lid inlaid with leather, 24-kt. gold stamped in a design created expressly for the Frederic Remington Revolver.

This beautiful hand engraving in period motif with silver inlays was done on a 3rd Generation Colt 1851 Navy with special cut down barrel and on this 1860 Army, by Conrad Anderson of Rocktree Ranch in Kingston, Idaho. Anderson gained fame as the engraver for the annual Roy Rogers Happy Trails Silver Screen Legend guns raffled each year to raise money for the Happy Trails Children's Foundation. Anderson also engraved all of Heritage Series Patersons for America Remembers.

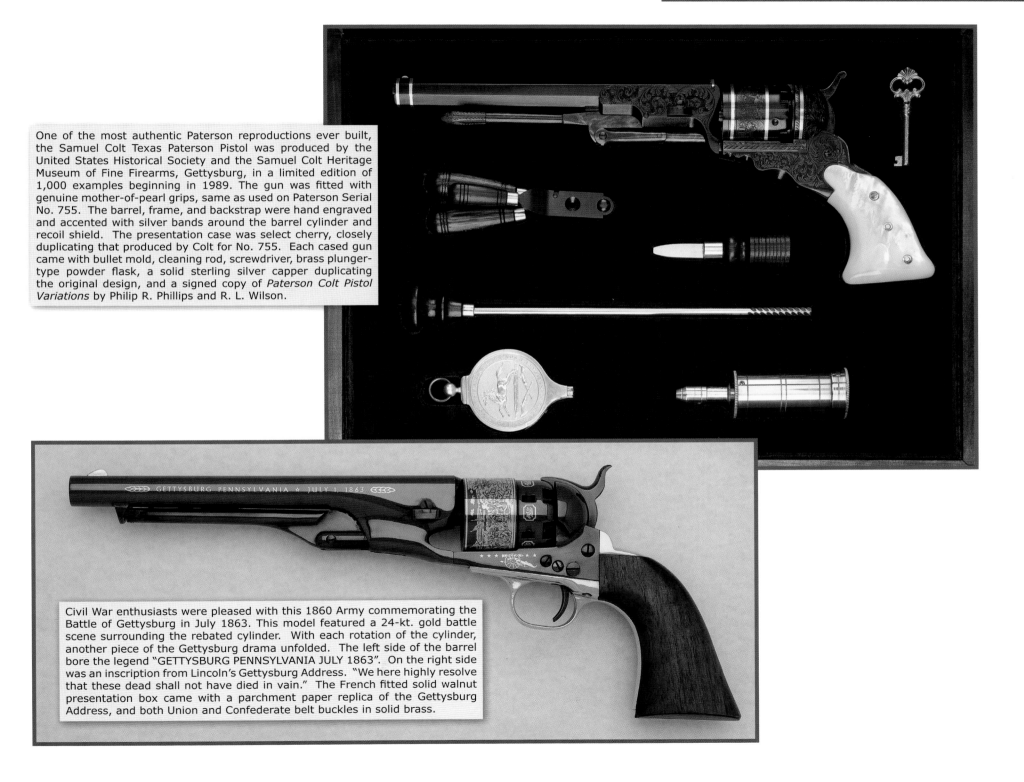

One of the most authentic Paterson reproductions ever built, the Samuel Colt Texas Paterson Pistol was produced by the United States Historical Society and the Samuel Colt Heritage Museum of Fine Firearms, Gettysburg, in a limited edition of 1,000 examples beginning in 1989. The gun was fitted with genuine mother-of-pearl grips, same as used on Paterson Serial No. 755. The barrel, frame, and backstrap were hand engraved and accented with silver bands around the barrel cylinder and recoil shield. The presentation case was select cherry, closely duplicating that produced by Colt for No. 755. Each cased gun came with bullet mold, cleaning rod, screwdriver, brass plunger-type powder flask, a solid sterling silver capper duplicating the original design, and a signed copy of *Paterson Colt Pistol Variations* by Philip R. Phillips and R. L. Wilson.

Civil War enthusiasts were pleased with this 1860 Army commemorating the Battle of Gettysburg in July 1863. This model featured a 24-kt. gold battle scene surrounding the rebated cylinder. With each rotation of the cylinder, another piece of the Gettysburg drama unfolded. The left side of the barrel bore the legend "GETTYSBURG PENNSYLVANIA JULY 1863". On the right side was an inscription from Lincoln's Gettysburg Address. "We here highly resolve that these dead shall not have died in vain." The French fitted solid walnut presentation box came with a parchment paper replica of the Gettysburg Address, and both Union and Confederate belt buckles in solid brass.

CHAPTER FOUR: Limited Edition and Collectable Models

Another exceptional commemorative from America Remembers was the massive and superbly embellished Walker Dragoon Sesquicentennial Tribute Revolver.

Pony Express 1851 Navy Revolver Commemorative was produced in a limited series by the United States Historical Society in 1992. The gun featured gold embellishments with frame and blued steel barrel decorated in 24-kt. gold. The etched images on the gold cylinder were based on drawings by William Jackson in the years before he became a famous photographer of the young West.

The United States Historical Society produced its own commemorative Robert E. Lee 1851 Navy in 1988. Unlike the Colt version done in 1971, the USHS model featured Florentine engraving in 24-kt. gold, based on the design presented to General Lee by Sam Colt prior to the Civil War. The USHS Navy also featured a special cylinder engraved in 24-kt. gold with five important symbols in the life of Robert E. Lee, including Lee's home in Arlington and the Lee family crest. A cast medallion of Lee was set into the left hand grip and a minted solid silver medallion with a bust of Lee on the obverse side and Lee atop his horse Traveler on the reverse was included in the French Fitted presentation case.

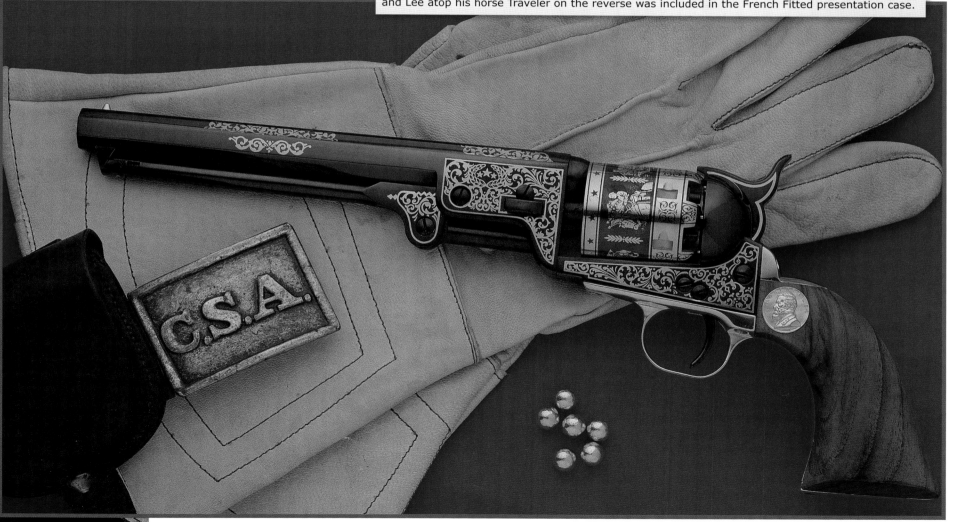

CHAPTER FOUR: Limited Edition and Collectable Models

Merrill Lindsay 1st Model Dragoon was produced by the United States Historical Society to commemorate the renowned firearms collector, historian and author. Cartoonist Charles Addams and Merrill Lindsay had been friends for many, many years. Therefore, when the U.S. Historical Society, headed by Bob Kline and Paul Warden, decided to produce a commemorative revolver honoring Lindsay, Addams was called upon to help. The details of the revolver had been suggested by Larry Wilson, based on the Whitneyville-Hartford Dragoon series. In fact, Larry made the first gun while at the Uberti factory, by going into the parts bins, and taking out a grip frame for a Walker and a cylinder and barrel for the 1st Model Dragoon, and putting them together as the initial prototype. The USHS very quickly had a new model of special-issue revolver.

The Whitneyville-Hartford Dragoon appeared again in 1998 as a special edition from America Remembers cased with a special singed and numbered edition of the original *Colt Blackpowder Reproductions & Replicas*. The standard Whitneyville-Hartford Dragoon remained in the Colt Blackpowder product line until 2002.

CHAPTER FOUR: Limited Edition and Collectable Models

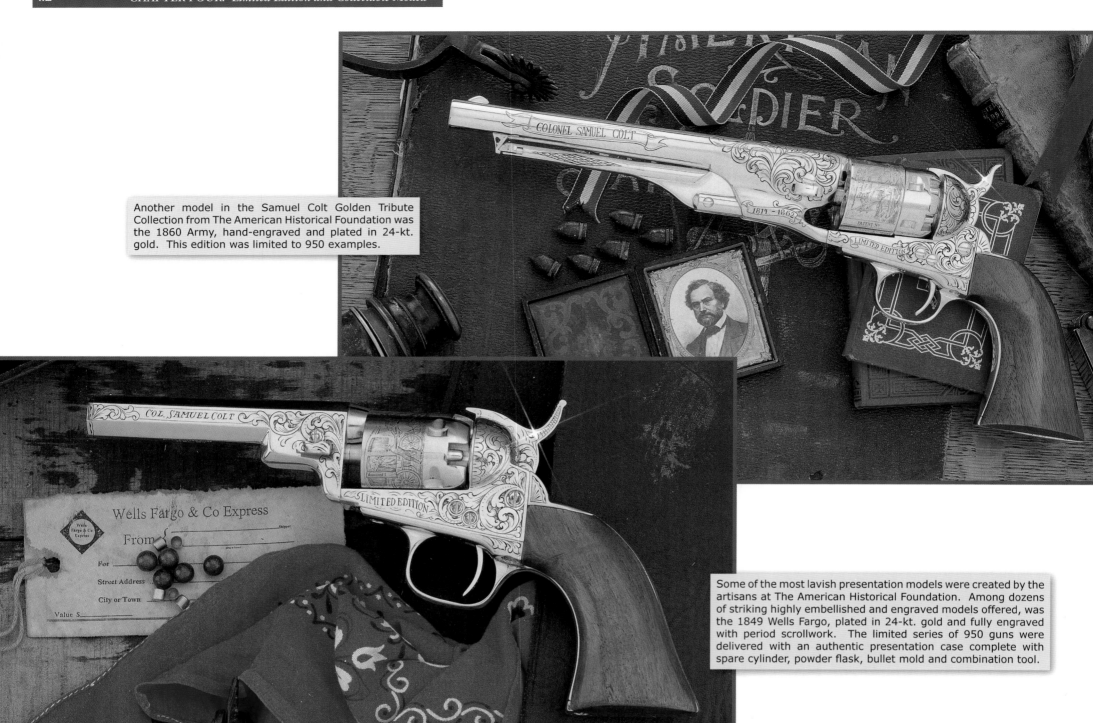

Another model in the Samuel Colt Golden Tribute Collection from The American Historical Foundation was the 1860 Army, hand-engraved and plated in 24-kt. gold. This edition was limited to 950 examples.

Some of the most lavish presentation models were created by the artisans at The American Historical Foundation. Among dozens of striking highly embellished and engraved models offered, was the 1849 Wells Fargo, plated in 24-kt. gold and fully engraved with period scrollwork. The limited series of 950 guns were delivered with an authentic presentation case complete with spare cylinder, powder flask, bullet mold and combination tool.

Black Powder Revolvers - Reproductions & Replicas

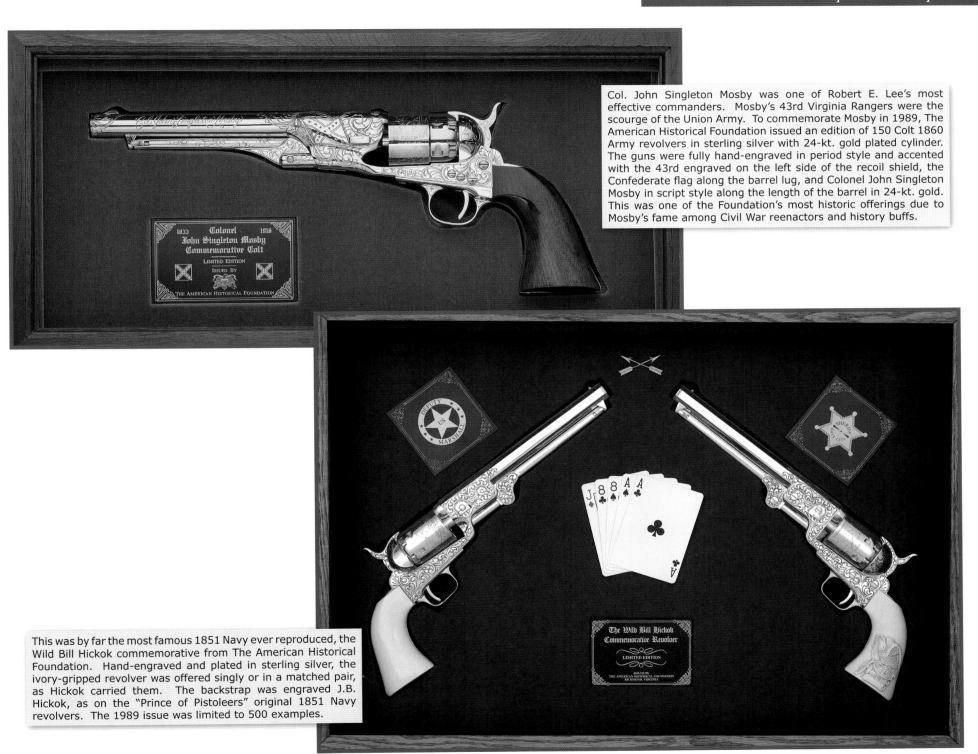

Col. John Singleton Mosby was one of Robert E. Lee's most effective commanders. Mosby's 43rd Virginia Rangers were the scourge of the Union Army. To commemorate Mosby in 1989, The American Historical Foundation issued an edition of 150 Colt 1860 Army revolvers in sterling silver with 24-kt. gold plated cylinder. The guns were fully hand-engraved in period style and accented with the 43rd engraved on the left side of the recoil shield, the Confederate flag along the barrel lug, and Colonel John Singleton Mosby in script style along the length of the barrel in 24-kt. gold. This was one of the Foundation's most historic offerings due to Mosby's fame among Civil War reenactors and history buffs.

This was by far the most famous 1851 Navy ever reproduced, the Wild Bill Hickok commemorative from The American Historical Foundation. Hand-engraved and plated in sterling silver, the ivory-gripped revolver was offered singly or in a matched pair, as Hickok carried them. The backstrap was engraved J.B. Hickok, as on the "Prince of Pistoleers" original 1851 Navy revolvers. The 1989 issue was limited to 500 examples.

On the 250th Anniversary of Commodore John Paul Jones, America Remembers produced another superb 1851 Navy. By this point in the company's history (America Remembers--United States Society of Arms & Armour, having become independent of the United States Historical Society), the special edition black powder arms were becoming more art objects than guns, and each new model was more ornate than the last. The 24-kt. gold and nickel decorations on the John Paul Jones created a magnificent contrast against the polished, blue steel finish. This model was limited to only 500 examples, delivered in an oak, glass-topped, French fitted display case.

The America Remembers Texas Ranger Colt Dragoon featured a silver cylinder engraved with the battle scene between Col. Jack Hays & Comanche Indians, as was depicted on the original Dragoon presented to Texas Ranger Ben McCulloch by Samuel Colt in 1848. The McCulloch 1st Model is the earliest surviving specimen of an engraved presentation Dragoon. The left grip on the commemorative features a sterling silver medallion inscribed with the Texas Ranger motto: "Free as the breeze, Swift as a mustang, Tough as a cactus." The Dragoon came in a presentation case with Colt Signature Series powder flask, combination tool, and bullet mold.

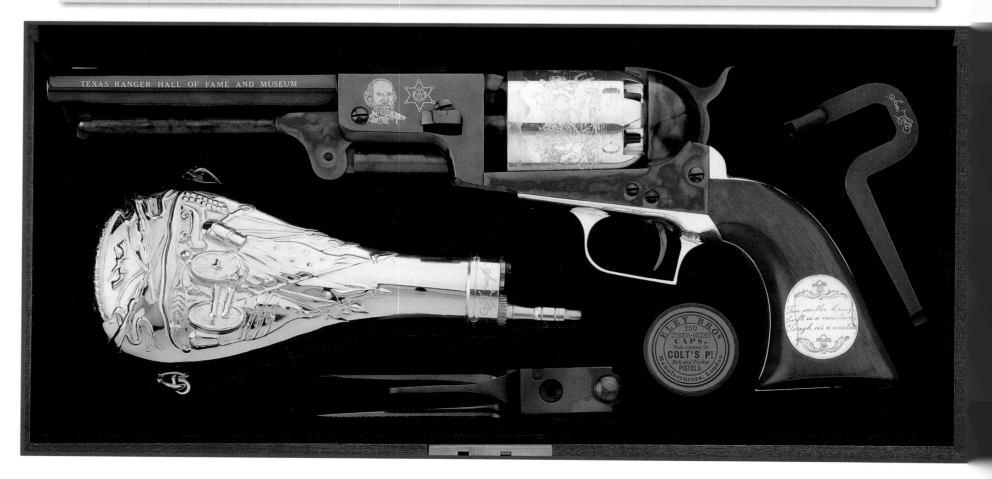

CHAPTER FOUR: Limited Edition and Collectable Models

In March 1990 The American Historical Foundation produced the largest commemorative pistol in its history, The Samuel Colt Golden Tribute Model 1847 Walker. Limited to 950 examples, this impressive 24-kt. gold plated revolver was hand-engraved in classic period scrollwork. Although none of the original 1100 Walkers are known to have been engraved, the design on the Golden Tribute Walker befits a gun of this caliber.

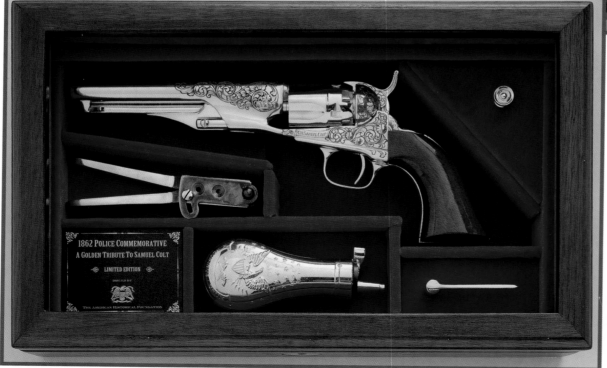

One of the most attractive of all Golden Tribute models was the 1862 Pocket Police. The superbly hand-engraved Colt was produced by Aldo Uberti and detailed with an authentic 19th century scroll pattern. No two revolvers were exactly alike in the series of 950 examples.

Black Powder Revolvers - Reproductions & Replicas

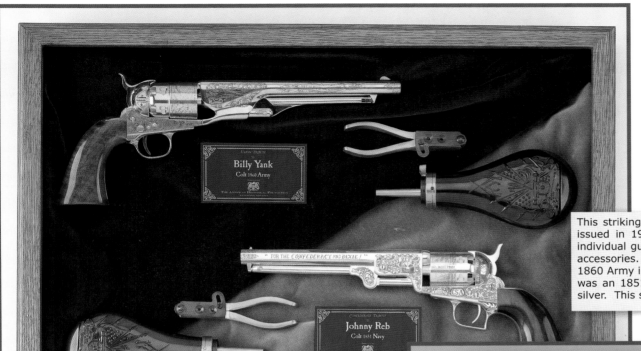

This striking pair of Union and Confederate commemoratives was issued in 1987 by The American Historical Foundation as either individual guns with flask and bullet mold, or as a cased pair with accessories. The Union Tribute to "Billy Yank" was a fully engraved 1860 Army in 24-kt gold. The Confederate Tribute to "Johnny Reb" was an 1851 Navy hand-engraved and plated in genuine sterling silver. This series was limited to only 250 of each.

Good as gold might have been the slogan for Sam Colt's first successful revolver, the No 5. Paterson Holster Model. The Texas Commemorative Paterson was superbly hand-engraved and plated in 24-kt. gold. The glass lid presentation box held the pistol, spare cylinder, powder charger, circular capper, combination tool, and cleaning rod. This series from The American Historical Foundation was limited to 950 guns.

CHAPTER FOUR: Limited Edition and Collectable Models

On the occasion of the 125th Anniversary of the American Civil War, The American Historical Foundation issued a pair of fully engraved Dragoons, a 3rd Model Union in 24-kt. gold and 2nd Model Confederate Dragoon in sterling silver.

In keeping with the North South theme of Civil War commemorative pistols, The American Historical Foundation 3rd Model Dragoon set was designed in contrasting blue and antique silver finish. Both models were hand engraved in a series of 250 each, beginning in March 1993.

Black Powder Revolvers - Reproductions & Replicas

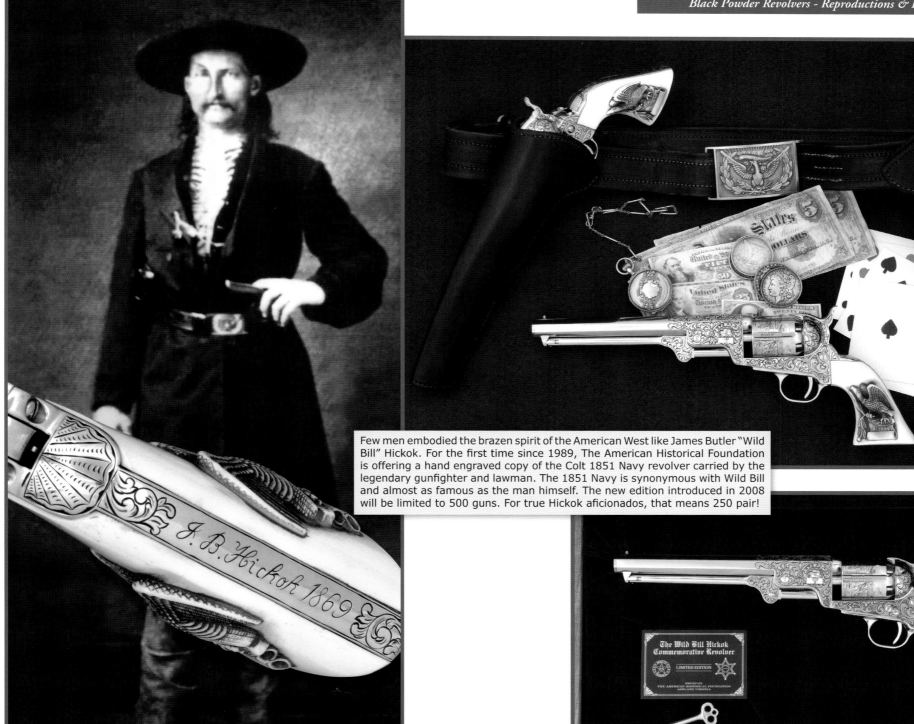

Few men embodied the brazen spirit of the American West like James Butler "Wild Bill" Hickok. For the first time since 1989, The American Historical Foundation is offering a hand engraved copy of the Colt 1851 Navy revolver carried by the legendary gunfighter and lawman. The 1851 Navy is synonymous with Wild Bill and almost as famous as the man himself. The new edition introduced in 2008 will be limited to 500 guns. For true Hickok aficionados, that means 250 pair!

CHAPTER FOUR: Limited Edition and Collectable Models

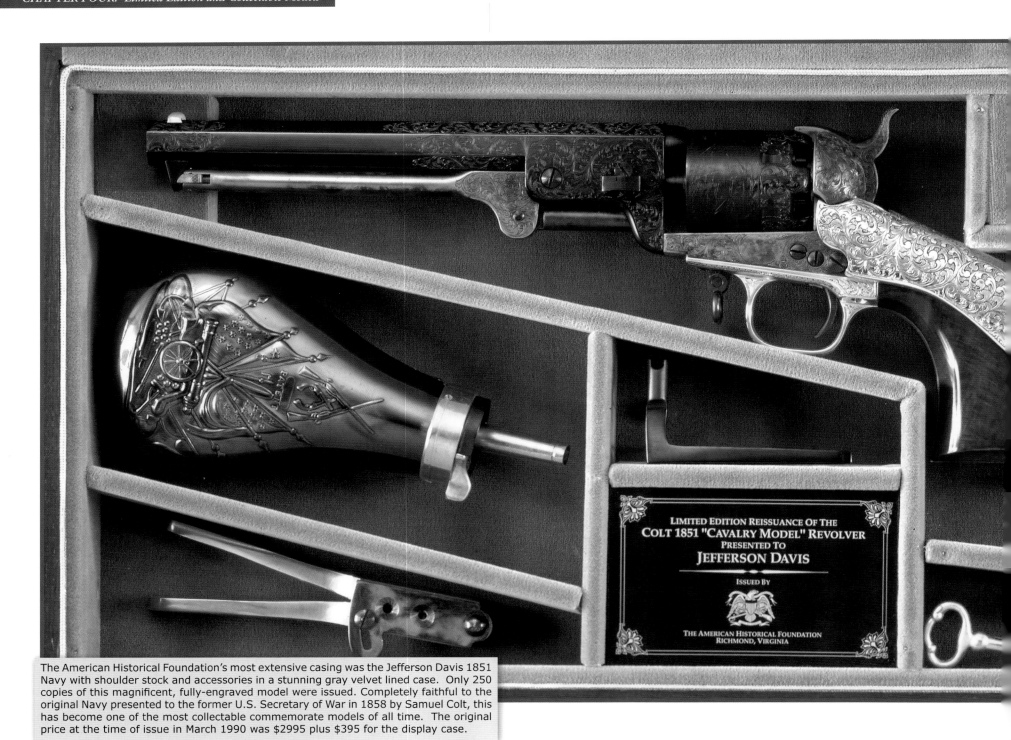

The American Historical Foundation's most extensive casing was the Jefferson Davis 1851 Navy with shoulder stock and accessories in a stunning gray velvet lined case. Only 250 copies of this magnificent, fully-engraved model were issued. Completely faithful to the original Navy presented to the former U.S. Secretary of War in 1858 by Samuel Colt, this has become one of the most collectable commemorate models of all time. The original price at the time of issue in March 1990 was $2995 plus $395 for the display case.

Black Powder Revolvers - Reproductions & Replicas 121

CHAPTER FIVE: *Shooting & Maintaining Black Powder Pistols*

> click...Click...CLICK. The hammer locks back. Sight, squeeze. Boom! A low, deep, thundering roar, a billow of smoke and a plume of flame. You have fired your first round in one of Samuel Colt's legendary blackpowder percussion revolvers. Shooting, as you have known it, will never be the same.

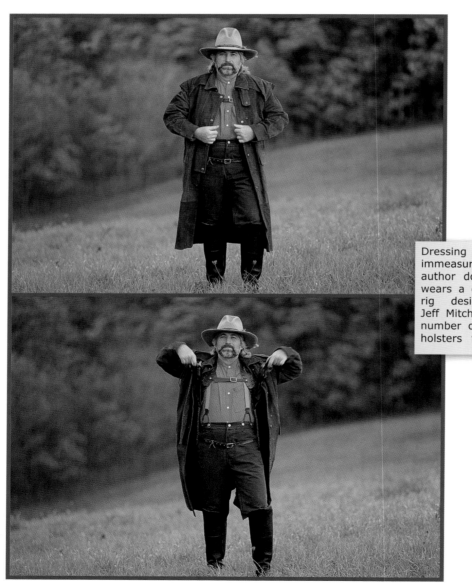

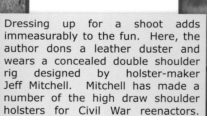

Dressing up for a shoot adds immeasurably to the fun. Here, the author dons a leather duster and wears a concealed double shoulder rig designed by holster-maker Jeff Mitchell. Mitchell has made a number of the high draw shoulder holsters for Civil War reenactors.

Whether you have chosen a hand-filling 1847 Walker Dragoon, a slender 1851 Navy, or one of Colt's lightweight .31 caliber Pocket Pistols, the end result is always the same – smoke, flames, and a throaty bawl unlike the report of any modern-day smokeless powder cartridge revolver. It is both the most appealing and heinous trait of the cap and

The design is based on one believed to have been worn by members of Col. John Singleton Mosby's 43rd Virginia Battalion of Partizan Rangers. While Civil War shoulder holsters are more likely a bit of fiction, it is surprising how quickly the guns clear leather with this rig.

Shooting & Maintaining Black Powder Pistols

Better Than James Butler Hickok Had It

124　CHAPTER FIVE: *Shooting & Maintaining Black Powder Pistols*

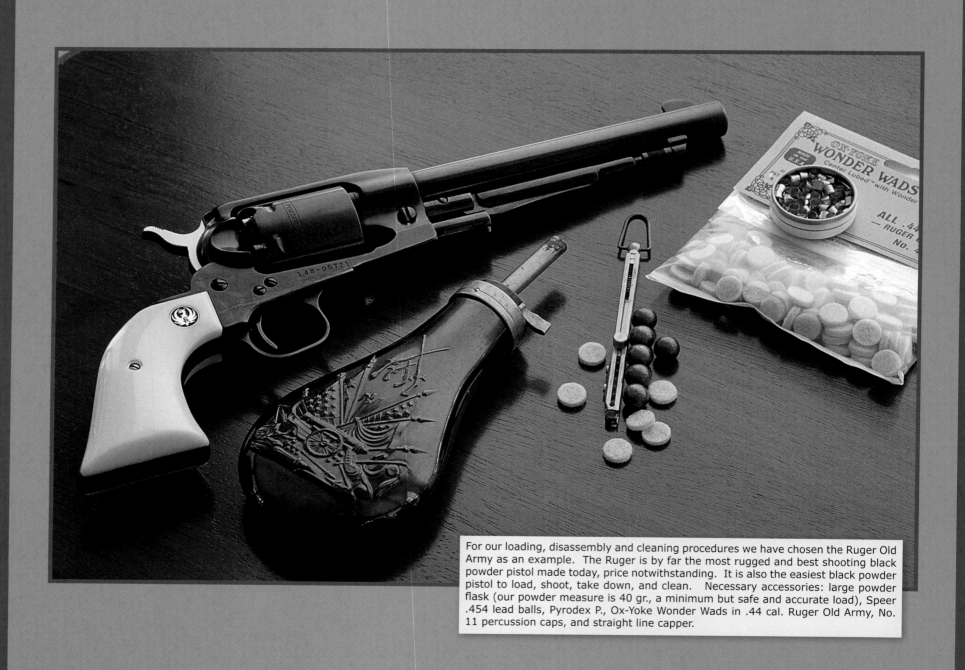

For our loading, disassembly and cleaning procedures we have chosen the Ruger Old Army as an example. The Ruger is by far the most rugged and best shooting black powder pistol made today, price notwithstanding. It is also the easiest black powder pistol to load, shoot, take down, and clean. Necessary accessories: large powder flask (our powder measure is 40 gr., a minimum but safe and accurate load), Speer .454 lead balls, Pyrodex P., Ox-Yoke Wonder Wads in .44 cal. Ruger Old Army, No. 11 percussion caps, and straight line capper.

ball pistol; the visual and aesthetic exhilaration of discharging a relic of American history, and the drudge of cleaning up afterward.

In the Old West, the West before Colt introduced the cartridge-firing Single Action Army, black powder revolvers reigned for a period of slightly more than 30 years, in calibers varying from .31 to .44, and in sizes small enough to conceal in one's pocket and large enough to strike fear into the hearts of men who found themselves staring down the barrel of a Dragoon.

The Colt 1851 Navy, 1860 Army and 1861 Navy were the most widely produced and carried sidearms of the Civil War, and in the late 1860s, the guns used in the settling (or unsettling, as the case may be) of the American West, along with Colt's Pocket model revolvers, and early cartridge conversions.

To the enthusiast-collector of these vintage repeaters, and moreover, to collectors of contemporary cartridge-firing pistols, the difficulty of loading and caring for cap and ball revolvers has always come as a surprise.

The loading process, however mastered, has always been a slow and arduous one requiring four steps, three if one was in a hurry, and perhaps a bit foolhardy.

First, the charging of each cylinder chamber with the proper amount of powder, then the loading of each successive chamber with a lead ball, followed by a covering of grease to lubricate the barrel and prevent the possibility of a chain fire (powder in an adjacent chamber accidentally being ignited, often with disastrous consequences), and lastly, the application of percussion caps.

If one could load each cylinder in the time it took to read that description of the process, they would have been considered quite proficient. With practice one can learn to do it. Now imagine loading this revolver in the heat of battle, surrounded by the roar of cannon, pistols, and muskets, thick clouds of gunsmoke, and the knowledge that you are only six rounds away from almost certain death. A bit melodramatic perhaps, but not far from the truth. Even soldiers who carried several guns and extra loaded cylinders had limited firepower before needing to reload or clean a fouled pistol. The cap and ball revolver was a great advance over the single-shot percussion pistol, but it was far from a perfect sidearm.

Such shortcomings provide the mystique which has surrounded these archaic 19th century handguns, and their popularity today among 21st century collectors and shooters.

From the early Texas Patersons to the refined 1862 Pocket Police and Pocket Model of Navy caliber, each Colt pistol, and at the same time, every Remington, LeMat, Starr, and Confederate model from the Civil War era, possess an elegance of design and function that far outshines the utilitarian paradigm of today's cartridge revolvers and semi-automatic pistols. In short, to those who look beyond the intended purpose of these legendary sidearms, black powder revolvers bear a certain cachet only realized with the passing of time. They have become a form of functional art, much in the same way as automobiles produced in the first half of the 20th century. We regard these icons from the past with perhaps undue reverence, but we do so nevertheless, and within the firearms industry this regard for 19th century percussion revolvers has brought about a renaissance in the manufacturing of 20th and 21st century replicas. Nothing else, however, has changed since the days of Beauregard's bombardment of Fort Sumter, the Battle of Little Round Top, and the Western Expansion of the 1870s. A percussion revolver built in 1848, or one produced in 2008, is no different in loading, firing, or cleaning. History then, is the tutor, and we are the students of tradition.

CHAPTER FIVE: Shooting & Maintaining Black Powder Pistols

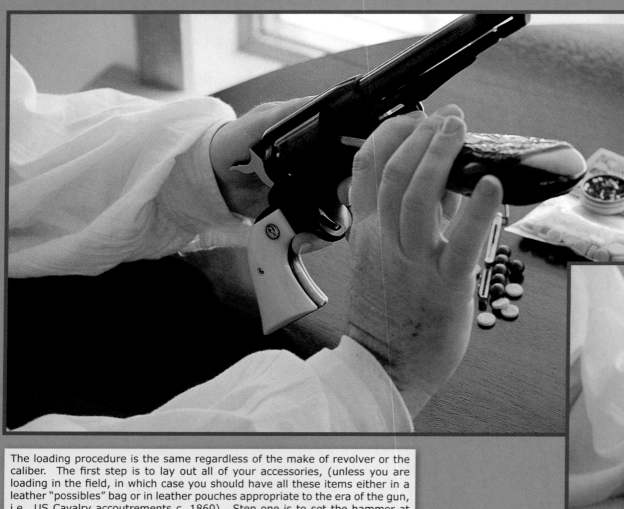

The loading procedure is the same regardless of the make of revolver or the caliber. The first step is to lay out all of your accessories, (unless you are loading in the field, in which case you should have all these items either in a leather "possibles" bag or in leather pouches appropriate to the era of the gun, i.e., US Cavalry accoutrements c. 1860). Step one is to set the hammer at half cock so the cylinder rotates freely. Begin by pouring a measured amount of powder into each chamber. Safety tip: We recommend following step two for each chamber to avoid the possibility of charging the same one twice.

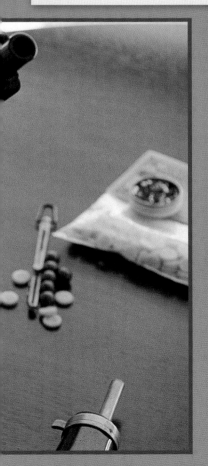

Step two. Again this is a personal preference, but we prefer to use the Ox-Yoke Wonder Wads for all shooting situations. The pre-lubed felt wads cover the powder charge and provide a solid seat for the round. By placing a wad over each chamber as the powder charge is dispensed, it becomes very obvious when you have all five (or six) chambers charged.

LOADING

The advent of the Ox-Yoke Wonder Wad (available in all calibers) has greatly improved both the loading process and the safety of black powder arms. The Wonder Wad, a felt pad saturated with lube, replaces grease in the loading procedure, and rather than being the third step becomes the second, providing both a lubricant for the barrel and a seat over the powder charge for the bullet (thus the etymology of the name wad). The use of Wonder Wads results in a cleaner, more consistent discharge, allowing more rounds to be fired between takedown and cleaning.

Revolver grease is still commonly used, and much less expensive than Wonder Wads, even more so if you use Crisco or lard, the latter taking one back to the 19th century. The Ox-Yoke product, however, is far easier to manage in the field, cleaner, and aesthetically more pleasing, as each round is visible in the cylinder, rather than a blot of white gunk.

There are arguments both for and against Wonder Wads, the dissenting opinion being that the lube is behind the round rather than in front of it. In addition, the wad is being sent down range as well, and at the instant of ignition is a buffer between the powder and the bullet. In our experience this has never proven to be a problem, but every shooter should draw their own conclusion by testing both methods. Most gun shops carry a variety of commercial gun lubes as well as Ox-Yoke products.

The first step in loading remains the same regardless of your choice of lubricants. With the hammer at half cock (to allow free rotation of the cylinder) the powder charge is dispensed into each chamber from either an adjustable powder measure or a brass or pewter flask fitted with a pour spout. Powder flasks in various sizes and with interchangeable threaded spouts ranging from 12 to 60 grains are available from virtually every major manufacturer and retailer of black powder firearms.

It is important to note that all calibers are not created equal. The powder charge for a .44 Army is not the same as a .44 Dragoon, nor is that the same as a .44 Walker or Ruger Old Army. Caliber is not the determining factor! Each pistol should come with a manufacturer's suggested powder load listed in the instruction handbook. If your gun does not have one, refer to the table of recommended loads in this chapter as a conservative starting point.

The first and foremost rule is load light. You can always work your way up. The consequences of over charging can be damaging if not down right dangerous. And never, ever use modern smokeless powder in a black powder firearm! Goex FFg, FFFg, Pyrodex P, Pyrodex pellets, Goex Pinnacle E-Z Loads and other commercial black powder substitutes are the only powders recommended. ~

CHAPTER FIVE: *Shooting & Maintaining Black Powder Pistols*

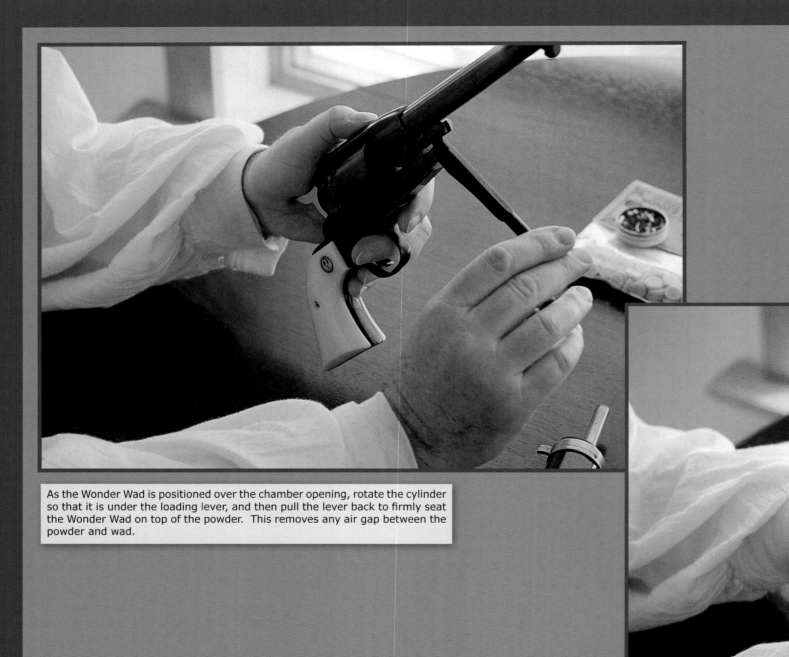

As the Wonder Wad is positioned over the chamber opening, rotate the cylinder so that it is under the loading lever, and then pull the lever back to firmly seat the Wonder Wad on top of the powder. This removes any air gap between the powder and wad.

POWDER, WAD, BULLET, CAP

After each chamber is charged, immediately seat a Wonder Wad, press it down on top of the powder charge with the loading lever, and then move on to the next chamber. This is a precaution to avoid accidentally charging the same chamber twice. (Paterson revolvers come with a five-spout charger that loads all five chambers at once.) If Wonder Wads are not being used, then seat a bullet over the chamber and ram it onto the powder before proceeding. Note that the top of the bullet should always be below the edge of the cylinder. If it extends above the edge, lead will be shaved off as the cylinder rotates behind the barrel and the gun more than likely will jam. If this happens, the barrel and cylinder must be removed to clear it, so make sure each round is properly seated. If you jam up a Remington in this fashion the problem is more difficult to correct.

After all six chambers (or five, depending on the gun and or your loading preference, i.e., leaving one chamber empty as a safety for the hammer), have been charged, the gun is now ready for the percussion caps.

The most common percussion cap sizes for revolvers are No. 10 and No. 11. Most often No. 11 will fit just about every revolver on the market. Occasionally, a No. 10 will fit better. A snug fit over the nipple is essential. Loose fitting caps on unfired chambers can fall off from recoil, or fired caps can fall into the hammer recess causing a difficult jam, and once again disassembly of the revolver to clear. A gun with loaded and capped chambers is not one you want to take apart unless there is no other option.

Once the cylinder is capped you have a fully loaded and deadly weapon. If the gun is to be holstered, either lower the hammer onto an empty chamber or let it rest on the stop between the chambers. This requires some practice with an unloaded gun to master the skill of setting the hammer in this position. It is not, however, a safety. Even on Remingtons and Ruger Old Armys, which have a notched hammer rest between chambers, this is not considered a safety. The only true safety is an empty chamber. ತ

Place a lead ball (again you have a choice between round balls or conical, our personal preference being round) over a chamber and rotate it around to the loading lever, firmly seating the round in the cylinder. It should be flush with the lip of the cylinder and not protrude. Rounds should also be firmly seated to ensure that no gap exists between the powder charge, wad, and lead ball.

130 CHAPTER FIVE: *Shooting & Maintaining Black Powder Pistols*

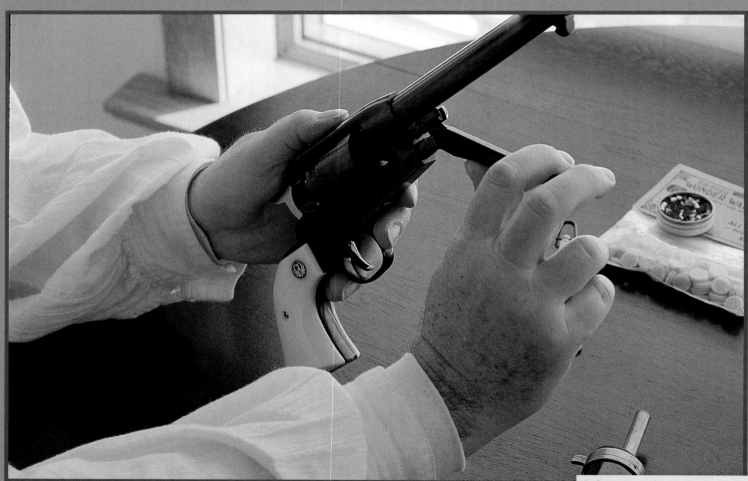

Safety tip. If you are shooting immediately after loading, charging all six chambers is considered a safe practice, however, since there are no safeties with black powder revolvers, if you are going to holster and carry your revolver, load only five, so that the hammer rests on an empty chamber. (On five-shot guns, load only four chambers).

The final step is to place a percussion cap on the nipple of each chamber. This can be done by hand if you have good dexterity, by pinching a cap between the thumb and forefinger and then placing it on the nipple. An easier and quicker method is to use a capper. Simply press a cap on to the nipple, rotate the cylinder and repeat. Safety tip. Your gun is now ready to fire and should be handled carefully, particularly when resetting the hammer. As a safety precaution, the hammer can be lowered to rest between the chambers. When it is cocked again the cylinder will complete its proper rotation and be in line with the hammer. This is the only practical safety the gun provides aside from lowering the hammer onto an empty chamber. On the Ruger Old Army, this is an integral part of the gun's design, and the hammer rests in a groove between the chambers. The same also applies to a Remington-style black powder revolver. However, this is not considered a safety.

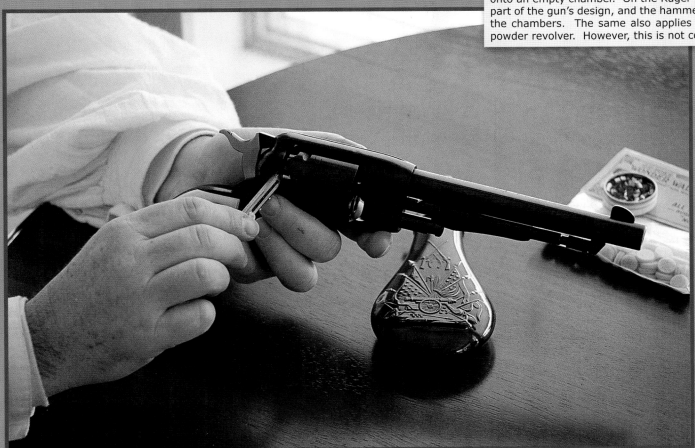

CHAPTER FIVE: Shooting & Maintaining Black Powder Pistols

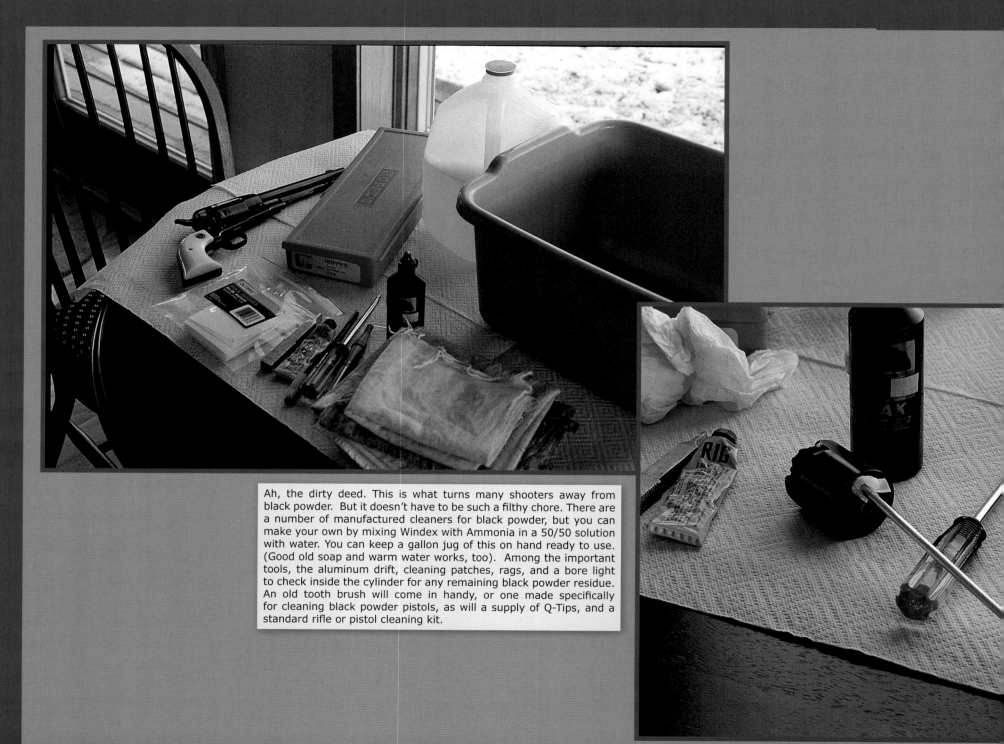

Ah, the dirty deed. This is what turns many shooters away from black powder. But it doesn't have to be such a filthy chore. There are a number of manufactured cleaners for black powder, but you can make your own by mixing Windex with Ammonia in a 50/50 solution with water. You can keep a gallon jug of this on hand ready to use. (Good old soap and warm water works, too). Among the important tools, the aluminum drift, cleaning patches, rags, and a bore light to check inside the cylinder for any remaining black powder residue. An old tooth brush will come in handy, or one made specifically for cleaning black powder pistols, as will a supply of Q-Tips, and a standard rifle or pistol cleaning kit.

AFTER THE FUN OF SHOOTING

For every action, there is an equal and opposite reaction. The gun fires and there is recoil. Even a .31 caliber Baby Dragoon will step back when the hammer drops. But that is only half of the "reaction." With the ignition of the powder charge a great cloud of smoke fills the air as the bullet leaves the barrel. What remains behind is a black powder residue coating the cylinder, center pin, barrel, frame, and your hand. By the end of six rounds you can see the buildup on the surface of the gun. After reloading the second or third time, the cylinder rotation becomes a bit stiffer, and by the fourth load the gun is fouled with powder. A black powder substitute like Pyrodex will give you a few additional loads because build up of fouling takes longer. A quick field strip – punching out the barrel wedge, removing the barrel and cylinder and giving the center pin, frame, and cylinder a quick wipe down with a powder solvent, (such as Birchwood Casey No. 77), followed by a light oiling of the center pin and exterior surfaces – will extend your shooting time. In a pinch, a wet cloth will clean off much of the exterior residue since black powder is water soluble. Sooner or later though, the gun is going to require a complete cleaning. This is usually around the same time you notice that your hands are dirtier than a chimney sweep's. Thus comes the counterpart to the thrill of shooting. The agony of cleaning.

Take heart, this is all part of the black powder experience. Like the lengthy loading process, cleaning is commensurately time intensive, but it doesn't have to be a begrudging chore. Our ancestors cleaned their guns whenever they were fired, and kept them up to snuff daily. Of course, their lives often depended upon it. For the black powder shooter the only life that depends upon a good cleaning is the gun's.

Nitrate powder residue and metallic fouling are harmful to your gun. Goex black powder is the most offensive, leaving an authentic 19th century mess on every part of the gun. Pyrodex P, and other synthetic brands of black powder create less residue and are easier to clean, but it's no picnic either. Black powder fouling is foul. A gun left unattended will suffer from corrosion, rust, and surface damage to the bluing, roll engraving and frame. This is particularly true of photoengraved pistols, (those with inexpensive laser etched scrollwork simulating hand engraving).

The black powder cleaning process varies according to the technique you prefer and the products used, from the simple, time-honored soap and warm water cleaning of all metal surfaces, (remember black powder is water soluble) to the use of high-tech products such as TDP Industries all-purpose SS-1 Cleaner Solvent, recommended by the U.S. International Muzzle Loading Team. Birchwood Casey also produces an excellent line of black powder solvents, and offers a complete Gun Care Guide free of charge.

Whatever you use, the most important thing is to get all the powder residue off the metal surfaces and clean the cylinder, chambers, nipples, and barrel bore thoroughly, then lightly oil all of the surfaces to prevent rust. Do it right, and do it every time. You can never be too meticulous. ❦

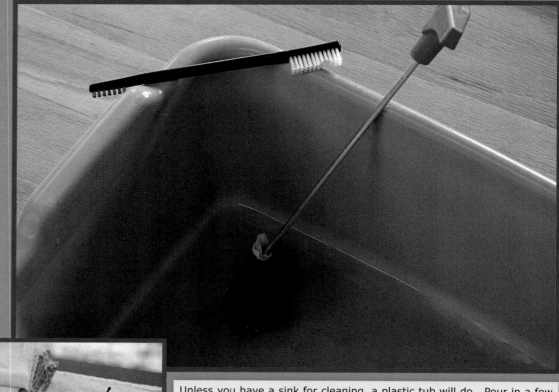

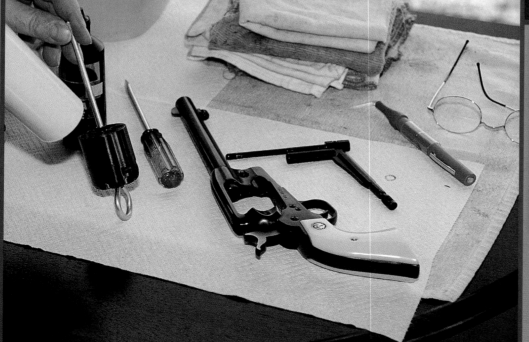

Unless you have a sink for cleaning, a plastic tub will do. Pour in a few inches of your cleaning solution and then drop in all of the parts. The cylinder needs the most attention, as each chamber must be thoroughly cleansed of all black powder residue, especially at the base of the nipple port. The back of the cylinder should be brushed around the nipples. After cleaning, a rinse in clear water will remove all of the cleanser. The cylinder then needs to be wiped out and dried. The same goes for the frame and barrel. Cleaning is the same as a conventional revolver, brushing out lead residue from the barrel, running patches down the barrel until they come out clean, etc. A lot of folks like to put the parts in the oven at a low temperature and bake out all of the moisture after cleaning. That works, but a 1500 watt hair dryer will do a great job and is easier to manage on a part-by-part basis. Some people have also been known to throw their entire gun, minus stocks, into the dishwasher. These people are known as single.

Black Powder Revolvers - Reproductions & Replicas

Clean up has been made easier with products such as Pyrodex EZ Clean and black powder solvents like Birchwood Casey's excellent No. 77, which virtually washes away fouling and residue.

Black powder substitutes like Pyrodex P provide optimal ignition and power with far less mess than black powder. Clean up is a little quicker because there is less fouling and residue. The same Pyrodex P substitute is also offered in pre-formed powder pellets in 30 gr. and 50 gr. sizes.

Barrels need the same attention and cleaning, regardless of model or takedown. The Ruger Old Army pictured, cleans the same as a conventional cartridge revolver. The entire gun should be lightly oiled. Break Free CLP is good for this job, as is WD-40, although there are those who disagree with using WD-40 because it dissipates. Gun oil, Break Free or WD-40, whichever you prefer for the gun, the center pin (cylinder shaft) threads should be lightly coated with RIG or similar grease.

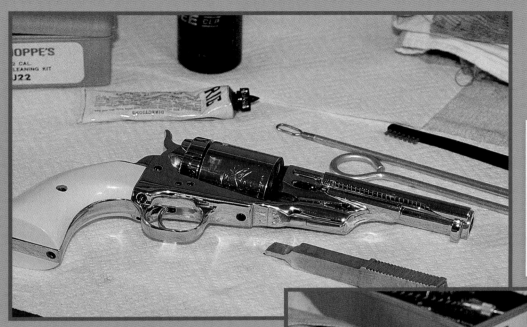

With a Colt or Colt-style pistol, such as this 1861 Navy custom made for the author by Dave Anderson, the traditional wedge must be driven through the barrel lug to begin disassembly. This is best done using a leather mallet to tap the wedge from its seated position. The leather mallet will not damage or mar the finish. The wedge (nitrite blued) can be seen protruding through the nickel finished barrel lug. Barrel removed from the center pin, the wedge can be seen in its retracted position in the lug. The wedge is held in place by a screw and should not be completely removed from the barrel. If you cannot find an aluminum drift, a plastic drift can be made from 3 inch square stock, cut approximately 6 inches in length with the tip cut into a drift. Ox-Yoke also manufactures a brass wedge drift.

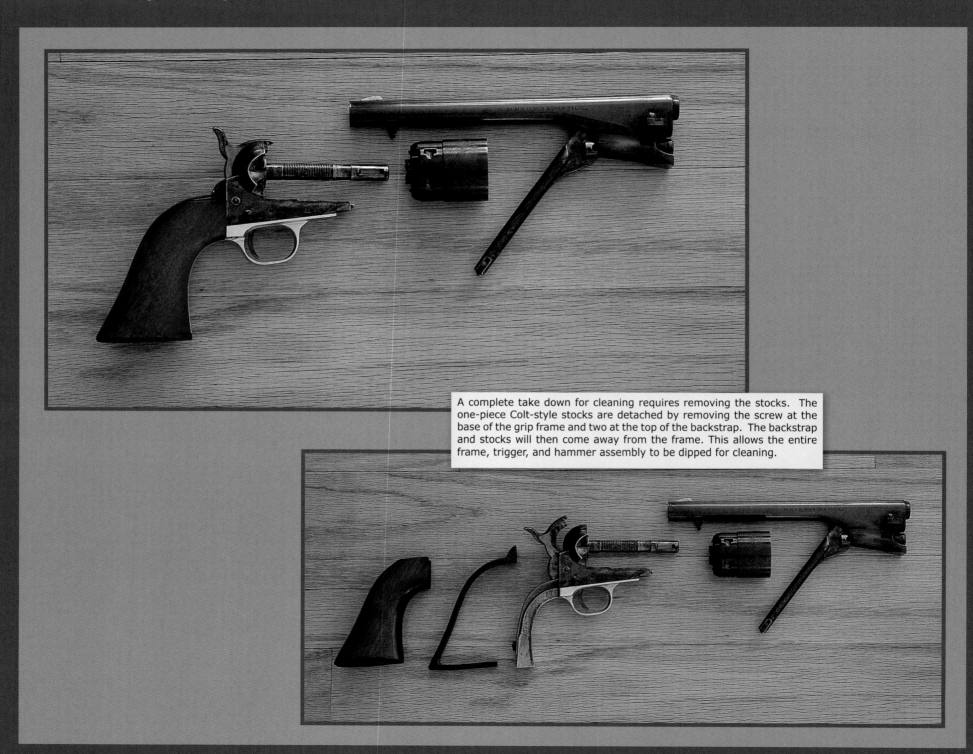

A complete take down for cleaning requires removing the stocks. The one-piece Colt-style stocks are detached by removing the screw at the base of the grip frame and two at the top of the backstrap. The backstrap and stocks will then come away from the frame. This allows the entire frame, trigger, and hammer assembly to be dipped for cleaning.

TAKE IT DOWN, DON'T BREAK IT DOWN

It's a gun. It's made of steel, powder explodes inside of it. How could you possibly hurt it? The steel parts, almost impossible. The bluing and surface finish, on the other hand, can be easily scratched or marred if improperly handled. The take down for cleaning, or field repair, is the number one culprit. Voice of experience here. The barrel wedge and barrel wedge slot are the most easily damaged pieces on your black power pistol.

Taking a Colt or Colt design revolver apart requires setting the hammer at half cock (as in loading) and then pressing the wedge through the barrel slot. This is often more difficult than it sounds. The wedge can be very firmly set. The best (and only sure way) to push the wedge trough is to tap it firmly with a leather mallet. Why leather? It's hard, but it won't harm the finish. If the wedge won't pull though, then it can be firmly tapped with an aluminum or hard plastic drift. (Never use a steel drift, it will damage the part.) Once the wedge is pulled through to the opposite side of the barrel, the loading lever is used to push the barrel assembly away from the cylinder by pressing it against the space between the chambers and then sliding the barrel off the cylinder arbor. The gun is now apart for general cleaning. The grips should also be removed to allow the frame, trigger, hammer, mainspring, etc. to be cleaned.

After the gun has been thoroughly cleansed of all fouling, completely oiled, and the center pin threads lightly greased, the pistol is reassembled and the wedge tapped back into place with the leather mallet. Here is where another problem can arise. The wedge should only be pushed far enough to engage the catch on the opposite side of the barrel. Tap it in too far and it cinches up the barrel and it will cause the cylinder to bind. If this happens, lightly tap the wedge back a bit until the cylinder rotates easily.

A properly cleaned and maintained black powder revolver will last a lifetime, actually, several.

The author test fires a Ruger Old Army loaded with 40 gr. of Pyrodex and a .455 lead ball. Recoil with this load is modest and accuracy is extremely good. The author was able to consistently shoot 1 1/2 inch to 2 inch groupings at 50 feet.

CHAPTER FIVE: Shooting & Maintaining Black Powder Pistols

GUNS LAID OUT BY CALIBER WITH FLASKS, WADS, BALLS

Guns and calibers. From the smallest to the largest pistols, our recommended loads are lighter than maximum for safety and accuracy. A big bang, more recoil, and lots of smoke looks impressive, but wears everything faster. We prefer a conservative load to reduce wear and tear on the pistol. Lighter loads can also provide more consistent accuracy. Between Goex FFFg and Pyrodex Pistol powder, and other black powder substitutes such as Goex Pinnacle, it becomes a matter of personal preference. The advantage of a substitute is a cleaner, less corrosive formula that allows more shooting before a gun becomes fouled.

Guns and Recommended Loads

All Colt and Remington, .31 cal: 12 gr. Goex FFFg, Wonder Wad, .315 ball. (Max load 13 gr.)

Colt Pocket Pistols, .36 cal: 18 gr. Goex FFFg, Wonder Wad, .375 ball. (Max load 20 gr.)

Colt 1851 Navy, 1861 Navy (similar copies) .36 cal: 22 gr. Goex FFFg, Wonder Wad, .375 ball. (Max load 25 gr.)

Colt 1860 Army, (Remington 1858 Army) .44 cal: 30 gr. Goex FFFg, Wonder Wad, .451 ball. (Max load 35 gr.)

Colt Dragoons (all but Walker Dragoon and Ruger Old Army) .44 cal: 40 gr. Goex FFFg, Wonder Wad, .451 ball. (Max load 50 gr.)

Walker .44 and Ruger Old Army: 40 gr. Goex FFFg, Wonder Wad, .455 ball. (Max load 60 gr.)

Pyrodex Data

.31 cal:	9 gr.	(Max load 10 gr.)
.36 cal Pocket:	15 gr.	(Max load 17 gr.)
.36 cal 1851/1861 Navy:	17 gr.	(Max load 20 gr.)
.44 cal. 1860 Army:	20 gr.	(Max load 28 gr.)
Dragoons:	35 gr.	(Max load 35 gr.)
Walker and Ruger Old Army:	40 gr.	(Max load 40 gr.)

Note: Although Pyrodex powder charges are actual weights not volume measures, the same spouts used for measuring Goex FFFg "by weight" can be used to measure Pyrodex "by volume." For example, in a Walker Colt the spout used to dispense 40 gr. of FFFg by weight, will deliver a correct measure of Pyrodex by volume.

Flasks and accessories courtesy of Taylor's & Co.

CHAPTER SIX: Behind The Scenes

A. Uberti S.r.l.'s new manufacturing plant in Serezzo, Italy.

Publisher S.P. Fjestad looks over one of the new CNC machines at A. Uberti's manufacturing facility in Serezzo, Italy.

The late Aldo Uberti in a photograph taken by the author in September 1997 at the Uberti factory. Aldo is holding the rough castings from which barrels and lugs are machined by CNC machines. Computer numerically controlled (CNC) machines have streamlined the manufacturing of black powder pistols.

Believe it or not these rough castings will become Colt frames after a trip through the CNC machines.

When film maker Sergio Lione and actor Clint Eastwood created the Spaghetti Western genre more than 40 years ago, little did anyone expect that Italy would become the world leader in the manufacturing of reproduction black powder revolvers. From *The Good, The Bad, and The Ugly*, to *The Outlaw Josey Wales*, right up to the Academy Award Winning film *Unforgiven*, which prominently featured a Starr double action black powder revolver in the hands of Eastwood, the allure of vintage pistols from the 19th century has captivated audiences and gun collectors the world over. It is perhaps quite fitting then, that the country which gave us the first great westerns to focus attention not only on the characters but on their guns, should produce the finest and most authentic reproduction 19th century pistols in the world.

It was with such knowledge that the author traveled to Italy in 1997, along with publisher S.P. Fjestad, renowned Colt historian and author R.L. Wilson, and sports car legend and gun enthusiast Luigi Chinetti, Jr., to visit the hallowed factory of Aldo Uberti & Co. in the historic gunmaking region of Gardone, and the modern manufacturing plant of Fratelli Pietta. These are Italy's two greatest and most revered suppliers of quality replica cap and ball revolvers. In 1999, after its acquisition by Beretta, A. Uberti S.r.l. moved into all-new manufacturing facilities in Serrezzo.

A. Uberti – Where it all began

As the late Val Forgett explained to the author in 1997, Aldo Uberti was the first gunmaker to successfully produce an authentic Colt black powder reproduction pistol. That was 50 years ago. In September 1997, we spoke with Aldo at the old

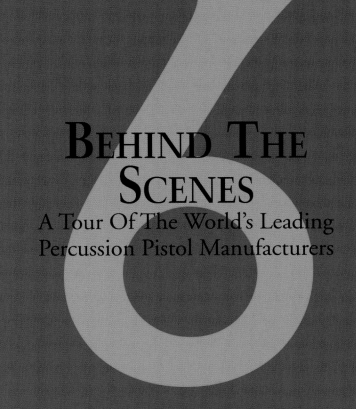

Behind The Scenes
A Tour Of The World's Leading Percussion Pistol Manufacturers

Gardone factory, and he recalled the early days of the industry. Sadly, Aldo passed away on March 21, 1998 at his home in Gardone Val Trompia, Italy, at age 74. For Aldo, it was a good life. The second youngest of six children (with two brothers and three sisters), Aldo's mother worked at the Armeria Bresciana factory, and was later employed at Beretta. His father had died when Aldo was three years old. He became a skilled mechanic, draftsman, and gunsmith, and by age 14 was working for Beretta, where he would remain for 21 years. The cradle of the Italian gun industry, Beretta was Aldo's training ground, where he mastered the art of gun making from the bottom up.

Aldo left Beretta in 1959, to set up A. Uberti & Co., working with Val Forgett (Navy Arms Co.) in creating what would become one of the hottest markets in firearms: replicas of 19th century cap and ball revolvers. The combination of Forgett and Uberti proved unbeatable, and in a matter of a few years both gentlemen were eminently successful.

The two collaborated on innumerable projects over the years, and can be credited, along with the late Turner Kirkland (founder-owner, Dixie Gun Works) with the launching of the replica business as a major segment of the firearms industry. Uberti & Co. supplied thousands of guns to director Sergio Lione, who visited the factory and residence, and was one of the gun maker's many admirers. Among others who shot Uberti guns were Clint Eastwood, Tom Selleck, Robert Mitchum, Kevin Costner, Kareem Abdul Jabar, Val Kilmer, Charlie Sheen, Mel Torme, Phil Spangenberger, and Sharon Stone. Virtually every Western movie made after 1960 featured Uberti-made firearms – among them *Young Guns, Silverado,* Kevin Costner's Academy Award winning *Dances with Wolves, Wyatt Earp, Tombstone,* and *Lonesome Dove* along with countless made for TV westerns and mini series. Uberti guns were also featured in the A&E special, *The Story of the Gun.*

"It is no exaggeration to say that Aldo was the Samuel Colt, D.B. Wesson, O.F. Winchester, and Eliphalet Remington of our time," says R. L. Wilson. "He mastered an art of making scores of models of guns, many of which had been out of production for over a century. He invented (and sometimes re-invented), processes and mechanisms, including building

A CNC machine at Uberti is set up to work on 1860 Army barrels, following a computer program to grind and mill rough castings. The nearly finished pieces then go on to final detailing, bluing, or plating.

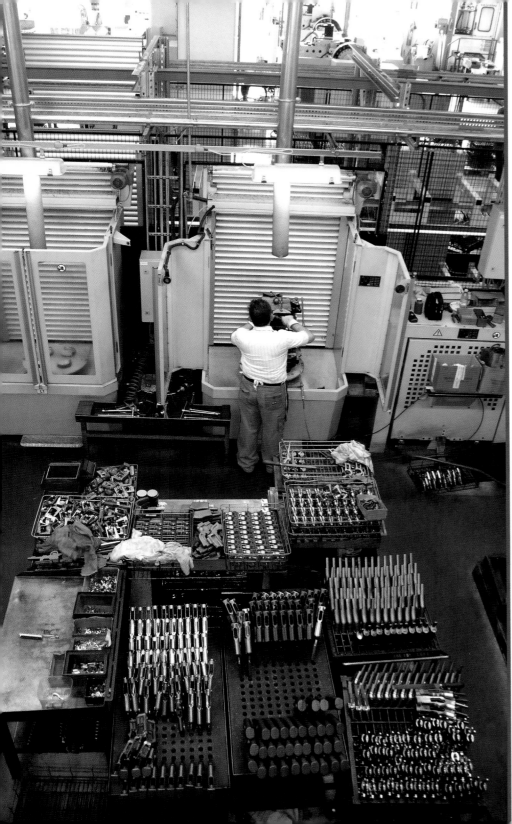

the Henry Rifle magazine tube, and designing an advanced safety system for the Single Action Army. All prototypes for Uberti replicas were built first by him, so he sorted out how each would be made, and was directly involved in the tooling. There wasn't a machine in the Uberti factory that the founder himself couldn't operate.

"When introduced to Bill Ruger, Sr., [many years ago] at a Shot Show, Ruger smiled and remarked: 'So you are the clever young man who designed the Single Action Army safety.' That design was patented by Aldo, and was the first advanced safety system for the time-honored mechanism."

"Daughter Maria Uberti remembers that her father did not know what 'bad' was. Despite not a few unpleasant experiences – in which he was imprisoned in a German concentration camp for supplying guns to partisans lifted from the Beretta factory in World War II, and occasionally having been cheated in a business deal – Aldo was so gentle and fair that he never recognized 'bad' in others. Many at the quiet funeral held in the Uberti home remarked of his purity."

"Few knew that Aldo was a talented artist, in addition to his known genius as a draftsman, machinist, inventor and master craftsman. Maria treasures a hand-carved ivory necklace made by her father while recuperating from lymphoma, in 1997. Aldo's brother was a gun engraver, a talent at which Aldo also excelled. With a file and a piece of wood, or ivory, it seemed as if Aldo could make anything. An Alpine skier and mountaineer, he had built spikes for his boots, and made his owning climbing pick. In his final months, Aldo was restoring a pair of elaborate Tschinke rifles, painstakingly replacing or repairing metal parts, replacing the missing forends and re-inlaying the stocks with ivory, bone, abalone, brass and pearl."

"Gunmakers, collectors, shooters, re-enactors and all who knew this gifted and gentle man will miss a true giant in the field of arms, whose creations have brought pleasure to millions, and whose product line helped save from wear and tear thousands of antique arms for present and future collectors and museums."

The manufacturing of a pistol even as technically primitive as a cap and ball revolver requires dozens of milling and grinding machines to produce all of the parts. In the new Uberti factory these machines help produce thousands of guns each year.

CHAPTER SIX: Behind The Scenes

The human element can never be entirely removed from manufacturing firearms. Detail polishing and finishing work are still done by skilled craftsmen. (photo by S.P. Fjestad)

Every Uberti pistol is hand assembled and inspected at the factory.

Walker Dragoon assemblies, frame, cylinder, barrel and lug, await final fitting, loading levers, backstraps and stocks.

The Gun In The Desk Drawer

In 1959 Aldo Uberti entered into a partnership with Gregorelli, a small company in Gardone, which was producing parts for Beretta. The very first 1851 Navy models were built by Gregorelli-Uberti for Navy Arms, and after an initial production of 2,000 revolvers, the demand rose to 1,000 Navy models a month. "This was more than we had expected," recalled Uberti in a 1997 interview with the author, "it was very successful."

Uberti added the 1851 Reb model in 1959 and two years later, he was the first to produce an 1858 Remington. Uberti went on to recreate virtually every cap and ball model ever produced by Colt.

It had all begun quite by chance when Val Forgett gave Gregorelli an original 1851 Navy and asked if it could be reproduced. "No one in the Gardone Valley wanted to do this. They laughed because here they were making shotguns," recalled Uberti. Gregorelli put the 1851 in his desk drawer, and that was where his partner, Aldo Uberti, found it in 1958. He asked, "What was this gun?" And Gregorelli told him about the American who wanted it reproduced. Uberti was intrigued by the idea, and very taken with the design of the Colt. He had never seen one before, and decided to take on the project. Uberti made all of the tooling to reproduce the 1851 Navy exactly as it had been made 107 years earlier in Hartford, Connecticut, by Samuel Colt. Today, Uberti remains the largest manufacturer of Colt and Remington reproductions in Italy, producing more than 15,000 black powder pistols each year, 80 percent of which are sold in America through gun stores and American distributors such as Taylor's & Co., Cimarron F.A. Co., Cabela's, and Dixie Gun Works.

The Fratelli Pietta – A Family of Gun Makers

Not far from Gardone, Brescia is the second seat of gun manufacturing in Italy, and home to Fratelli Pietta. Established in 1960, F.lli Pietta moved into its new, modern assembly plant in 1997, becoming one of the most technologically advanced black powder arms producers in the world. Today, with new expansions in the works, the Piettas have come to be known for recreating guns so elaborate and complex that no one else has ever attempted to make them.

Since the 1960s Pietta has played an important role in the world production of replicas. The company originally manufactured hunting guns in the variations of single barrel, side by side, and over and under. In 1964, Pietta branched out into the lucrative black powder arms field producing the Colt 1851 Navy. The large and immediate success of the Navy led to the introduction of other famous models, the 1860 Army, 1858 Remington New Model Army, the Spiller & Burr, the Paterson, the Dance Dragoon, Starr single and double action revolvers and the legendary LeMat, which was awarded the prestigious prize as "the best gun of the year" in 1985 by the American National Association of Federally Licensed Firearms Dealers.

With nearly half a century of experience, F.lli Pietta has developed its own style of manufacturing which combines modern machinery with old world craftsmanship. While high technology CNC machines produce parts with unerring accuracy, every gun is fitted, polished, and assembled in the time honored tradition of European and American gunmakers, by hand, one gun at a time.

Blued and finished Remington New Model Army revolvers await their cylinders on the way to final inspection.

Every gun begins here in the computer design studio were original firearms are plotted on a spectrograph and computer drawings are created for the intricate CNC tooling used to make each piece. The gun being designed is a reproduction of the 1858 Starr double action revolver. The Starr was used extensively by the Union Army during the Civil War, and was added to the unique line of 19th century reproductions manufactured by F.lli Pietta in 1998.

The company, which is operated today by founder Giuseppe Pietta and his sons, Alessandro and Alberto, utilizes the latest computer aided design (CAD CAM) and spectrographic equipment to accurately reproduce the most historic sidearms of the 19th century.

Among the finest examples of Pietta design and quality is the 1858 Starr double action and single action revolver, in both .36 and .44 caliber. Aside from the award winning LeMat, the Starr is the most complicated design ever undertaken by Pietta, and has become one of the most sought after guns for Civil War reenactments. This elaborate pistol took more than a year to develop and in 1998 brought the Pietta black powder line to a staggering 79 models, including a stunning series of fully hand-engraved models in the Colt, Remington, and LeMat line.

Pietta's commitment to producing quality arms at affordable prices has made their black powder line, sold by Cabela's, Navy Arms, Cimarron F.A. Co., Taylor's & Co., Dixie Gun Works, EMF, and Bass Pro Shops, one of the most popular in the United States.

The Piettas have ventured further into the history of the American Civil War than any other gunmaker, producing not only Colts, Remingtons, and LeMats, but lesser known and historically significant Confederate makes such as the 1862 Griswold and Gunnison, the 1862 J.H. Dance and Brothers .36 caliber revolver, and the 1862 Spiller & Burr .36 caliber repeater. Pietta is the only manufacturer today producing the unusual J.H. Dance revolver, sold though Dixie Gun Works. Pietta also sells a complete line of French fit display cases, compartment cases, powder flasks, bullet molds, and accessories for every caliber and make of revolver they manufacture.

Recreating The Unusual

There have been a handful of arms makers in Italy who specialized in the more eclectic of 19th century revolvers, and one of the more adventurous was Palmetto in Brescia which built one of the more uncommon Colt revolvers, the Roots pistol. Palmetto's guns were based on the most successful of the seven

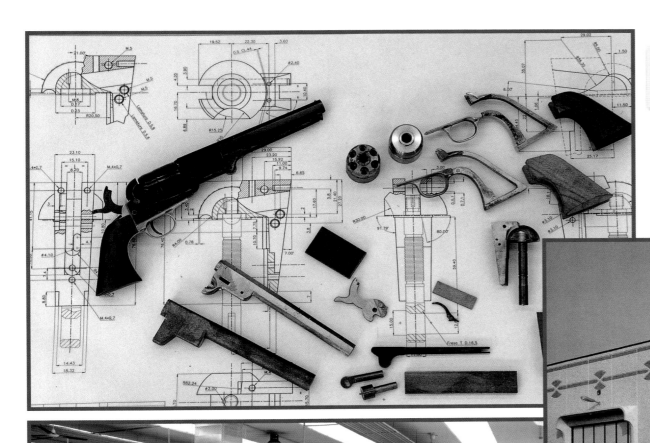

Piece by piece, from computer generated blue prints to tooling, Pietta creates every part for their revolvers. Shown are a finished 1851 Navy, the technical drawing, and comparisons of rough cast and finished pieces.

The Fratelli Pietta factory in Brescia, Italy, produces many of the world's finest black powder revolvers including examples of the LeMat and Starr, two of the most elaborate and complex firearm designs of the 19th century. (Photo courtesy F.LLI Pietta)

Pietta's high tech CNC shop is contrasted by the individual stations on the main floor (left) where components are hand-fitted and assembled, one gun at a time.

CHAPTER SIX: Behind The Scenes

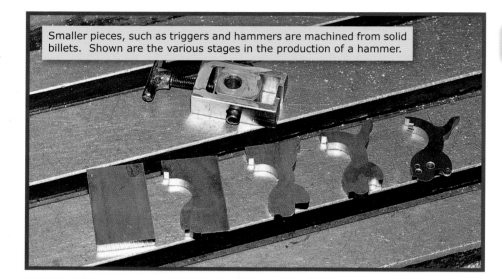

Smaller pieces, such as triggers and hammers are machined from solid billets. Shown are the various stages in the production of a hammer.

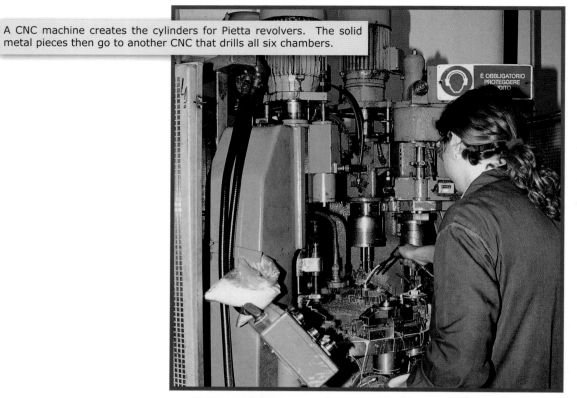

A CNC machine creates the cylinders for Pietta revolvers. The solid metal pieces then go to another CNC that drills all six chambers.

No one has yet created a machine that can assemble and fine tune a pistol like the skilled hands of a seasoned craftsman. At Pietta, each revolver is inspected, hand fitted, and then checked for cylinder rotation, hammer and trigger action. The guns are then given a final inspection before being packaged.

The author in 1997 with the Pietta family, the company founder Giuseppe Pietta and his sons Alessandro (left of author) and Alberto.

Black Powder Revolvers - Reproductions & Replicas   151

A 3rd Generation Colt was a Colt, mostly. Cast components in the white (unfinished) were produced to Colt Blackpowder Arms' specifications at the foundry in Italy and shipped to Colt for finishing, fit and assembly, case hardening, and bluing. Here a case of barrels arrives for inspection in 1997.

Each piece was hand polished to ensure a smooth finish up to Colt standards.

Back in 1997 polished barrels and frames awaited the next step in assembly at Colt Blackpowder Arms in Brooklyn, New York.

The Colt Formula

Bringing back the past is an art form that few had mastered as well as Colt Blackpowder Arms, which utilized the original Colt proprietary formula for case-hardening the frames, loading levers, hammers and recoil shields, and rendering that brilliantly marbled steel finish for which the original pistols have been acclaimed.

Case hardening is one of the most distinguishing characteristics separating Italian-made reproductions from those that were produced by the Colt Blackpowder Arms Company.

In 1997 Colt Blackpowder Chairman, Louis Imperato explained to the author that the case hardening formula, as the name implies, hardens the exterior surface of the metal, so that it wears better. "The colors created from traditional case hardening (a combination of super heating and rapidly cooling the metal parts which are packed in charcoal and bone meal)," said Imperato, "are deeper, with a swirling grey, brown, black effect, a much more elegant look than is achieved by the Italian cyanide process. Italian-made guns have more blues and grays in the finish, and not as much depth."

The charcoal and bone meal formula used by Colt was originated in England and used on Holland & Holland and Purdey shotguns. Colt adopted the British formula more than a century ago, and that is what Colt and Colt Blackpowder used.

Imperato noted that the exact mix of charcoal and bone meal and the specific temperatures vary depending upon the individual parts being hardened. "Each part needs its own recipe. A frame and a loading lever would be different than a hammer, for example. This is a costly process compared to cyanide dipping. The same is true of our bluing process, the finish is much deeper, but that isn't entirely bluing. Our chemicals are a little bit different, but it is more a result of the actual preparation of each gun before the bluing. We put a lot of time and effort into the polishing of each piece before it is blued. Polishing is an essential step. You have to create a highly polished finish but not loose the edges. You have to be careful not to round off an edge. If it's a square edge it has to remain square. If you look at a Colt trigger guard, for example, you'll find

From red hot to ice cold, the parts were dumped into chilled water in an explosion of fire and smoke. This was all part of Colt's case hardening process. In Italy, with few exceptions, a case hardened finish is created through a process which requires dipping the pieces in a cyanide solution. The finish is noticeably distinct from that of a Colt's color and brilliancy. This is part of what one paid for in a Colt Blackpowder pistol.

that there is a chamfered edge around the trigger guard itself. On most guns you'll find that the edges are rounded off from polishing. That's the beauty of a Colt, those sharp edges, the deep case hardened colors, the bright bluing, that's what makes a Colt a Colt." Today, those same attributes make the 3rd Generation Colt Blackpowder revolvers highly desirable among shooters and collectors alike.

A frame is inspected for finish before going on to the case hardening process.

Guns at Colt's were assembled and reassembled, fitted and refitted until they went together with exacting precision.

Root's pistol variations first patented in 1855 and featuring a solid frame and unique side mounted hammer. The Root pistols and revolving rifles became eminently popular during the War Between the States, particularly the latter which allowed a soldier to have a large caliber rifle, carbine, or shotgun with up to six shots. The pistols were built from 1855 through 1873 in seven different series variations chambered in either .28 or .31 caliber. Palmetto's copies were sold exclusively through Dixie Gun Works. Palmetto also built the Root Revolving rifle sold by Dixie, as well as the famous Civil War Model 1855 Dragoon Pistol with shoulder stock, a variety of Derringers, and the very rare Model 1861 Whitney revolver, all of which are only available on the secondary market.

Euroarms Italia (formerly Armi San Paolo) in Concesio, is another of Italy's oldest replica arms makers, and like Palmetto has built a reputation for manufacturing rare and unusual revolvers from the Civil War era. There most famous is the great Rogers & Spencer revolver which shares some of its appearance and history with the solid frame Whitney Second Model 3rd Type revolver originally built in Connecticut during the early 1860s. The Rogers & Spencer was produced in Willowvale, N.Y. between 1863 and 1865; however, the majority of the guns manufactured under Union contract were not delivered until after The War Between the States was over. Most became Army surplus guns in the 1870s where they saw considerable use on the frontier. The Euroarms model is an exacting, quality replica. In the past, Armi San Paolo also produced copies of the Griswold & Gunnison Navy model as well as Colt and Remington revolvers of the Civil War era.

One of Colt's master gunmakers files a barrel to ensure an exacting fit to the frame. This is what separated many Italian copies from guns bearing the Colt name and patents.

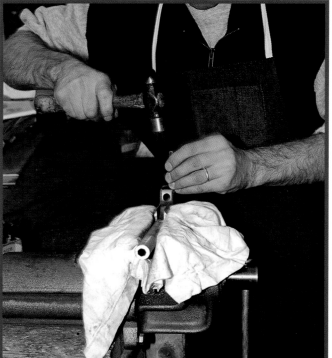

After a gun was perfectly fit, and only then, were serial numbers stamped by hand into the barrel and frame. The parts then moved on to their respective departments for bluing or case hardening.

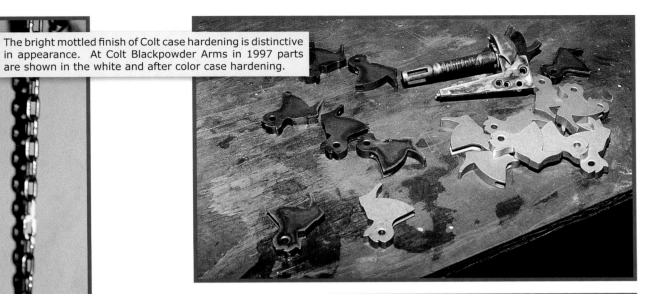

The bright mottled finish of Colt case hardening is distinctive in appearance. At Colt Blackpowder Arms in 1997 parts are shown in the white and after color case hardening.

Colt Blackpowder Arms used Colt's proprietary bluing formula which rendered a finish unmatched by most gunmakers in the world.

After polishing, bluing and color case hardening, serial numbered components were re-matched and sent to final fit and assembly at Colt Blackpowder Arms' Brooklyn assembly plant. This should answer once and for all the question of whether 3rd generation Colt's were made in America.

Select walnut stocks were cut to fit each Colt pistol.

Guns got the white glove treatment as they went to final assembly where each was hand-fitted and inspected.

The finest quality copy of the Rogers & Spencer is actually made in Oberndorf, Germany, by the legendary Feinwrkbau armory established in the 1950s. The Rogers & Spencer is the only percussion revolver produced by Feinwrkbau and is marketed by Davide Pedersoli.

There is indeed a pecking order within the Italian firearms industry, and Uberti, F.lli Pietta, Davide Pedersoli (which manufacturers earlier style single shot pistols and long arms) and ArmiSport di Chiappa (known for their fine reproductions of Winchester lever action rifles, historic French dueling pistols, Civil War muskets, Sharps rifles, and the Spencer carbine and musket) are at the very top. The great diversity of manufacturers, models, and prices available today, offer collectors and shooters the finest and most extensive series of historic 19th century firearms, since the 19th century.

Colt Blackpowder President Anthony Imperato didn't pack each gun personally, but he was always there to oversee the operation and ensure the quality and reputation of Colt Blackpowder pistols. A gun that bears the Colt name is one that was carefully built, not just the finish but the metal to metal fit, and the wood to metal fit.

CHAPTER SEVEN: Practical Percussion Pistol Shooting

We are all frustrated actors.

Through clouds of gunsmoke, Confederate troops defend their last position with pistol fire. This is the most exciting part of the reenactment. Crowds of spectators moving along to follow the battle are just out of camera range behind the scenes.

Wouldn't it have been nice to play a bit part in *Tombstone*, or *Deadwood*, or to have had a moment in front of the camera during the filming of *Gettysburg* or *Gods and Generals*? For a handful of Civil War reenactors and Cowboy Action Shooters, their 15 seconds of fame came and went as movie extras in both films and television shows, either charging through the battlefields of Gettysburg, or walking along the streets of Tombstone or Deadwood. For most of us though, playacting is just a fantasy.

Or is it?

7
Practical Percussion Pistol Shooting
Skin That Smokewagon!

Since the days when Val Forgett first introduced the reproduction Colt 1851 Navy, there have been Civil War reenactments. It was the Centennial Celebrations of the Civil War, from 1961 to 1965, that launched the entire replica arms industry. Today there are Civil War battle reenactments taking place almost every weekend somewhere in the country. In Pennsylvania, there are reenactments scheduled throughout the summer, and at Gettysburg, reenactors not only get to replay history, but do it on the very ground where it took place.

What attracts everyone from business executives, police officers, and artists, to auto mechanics and salesmen to don hot, heavy fabric uniforms and forsake all the comforts of the 21st century for a weekend encampment in 19th century tents, and food cooked over an open fire? Well, it isn't camping out, because at least twice a day they go into mock battle with guns blazing, pounding hooves, Rebel yells and the roar of cannon and either triumph or die in the field, depending upon the battle scenario being reenacted.

For most Civil War reenactors, it is the fun of dressing up in costume and playing a part that brings them into the hobby. Others take it more seriously, students of the Civil War, who adhere to every covenant of the era and spend

The troops fall in as Confederate infantry is heard to be in the vicinity. Reenactments, such as this one at Old Bedford Village, in Bedford, PA, follow an actual battle that took place during the Civil War.

their time in uniform as though reliving a page from American history. Either way, Civil War reenactors enjoy the art of entertaining onlookers with carefully staged battles taken from real events.

The chain of command is correct from generals to infantrymen and many reenactors have taken on the part of a specific historic character, which they play year in and year out like a recurring role in summer stock. There are literally hundreds of well known and little known reenactors who have mastered the characters of famous commanders and soldiers from both sides of the War Between the States, from Grant and Lee to George Armstrong Custer.

This is a pastime that always welcomes new members, and in recent years the enlistment roster has been growing as more and more Americans, young and old, men and women, find this hobby an entertaining and exciting way to not only spend a weekend but become a part of something bigger, American history.

A Confederate rifle volley opens the engagement between North and South.

CHAPTER SEVEN: Practical Percussion Pistol Shooting

At many reenactments cannon fire is one of the most exciting parts, both for the reenactors and spectators!

Confederate cannon and caisson move into position to answer Union cannon fire.

CHAPTER SEVEN: Practical Percussion Pistol Shooting

Carefully checking the corners, a Union Captain prepares to have his men advance.

Union infantry opens fire on a group of Confederate soldiers moving among buildings.

Southern troops advance across the field of battle with pistols blazing and the Rebel yell filling the air.

Union and Confederate troops unleash a volley of fire at each other in open field combat during a 2007 reenactment at Bedford Village in Bedford, PA.

CHAPTER SEVEN: Practical Percussion Pistol Shooting

Confederate troops prepare to move against a Union position as troopers with flag unfurled open fire on the distant Northerners.

Black Powder Revolvers - Reproductions & Replicas 169

CHAPTER SEVEN: Practical Percussion Pistol Shooting

Union and Confederate commanders on horseback plan their attack strategies at one of Beford Village's largest reenactments involving mounted cavalry.

CHAPTER SEVEN: Practical Percussion Pistol Shooting

The wide open battlefield at Bedford Village allows for large troop movements and cavalry charges. These are some of the most exciting moments in a reenactment, just as they are in a film.

The differences in Union and Confederate uniforms and equipage is evident in these two views of troops from both sides of the conflict.

A typical complement of side arms for a Civil War reenactment, an 1862 Pocket pistol, 3rd Model Dragoon, 1861 Navy with short barrel, 1860 Army with fluted cylinder, 1851 Navy, 1860 Army, 1858 Remington Navy, a Spiller and Burr, a Baby Dragoon, a pair of Griswold and Gunnison revolvers (one with stag stocks), an 1847 Walker, an 1860 Army, and a Pepperbox. All good shooters, most were produced by F.lli Pietta.

Soldiers and drummer boys take a break for a hand of cards during encampment at an Old Bedford Village Civil War reenactment.

One of the rare Root revolving rifles (left) that was manufactured by Palmetto and sold by Dixie Gun Works.

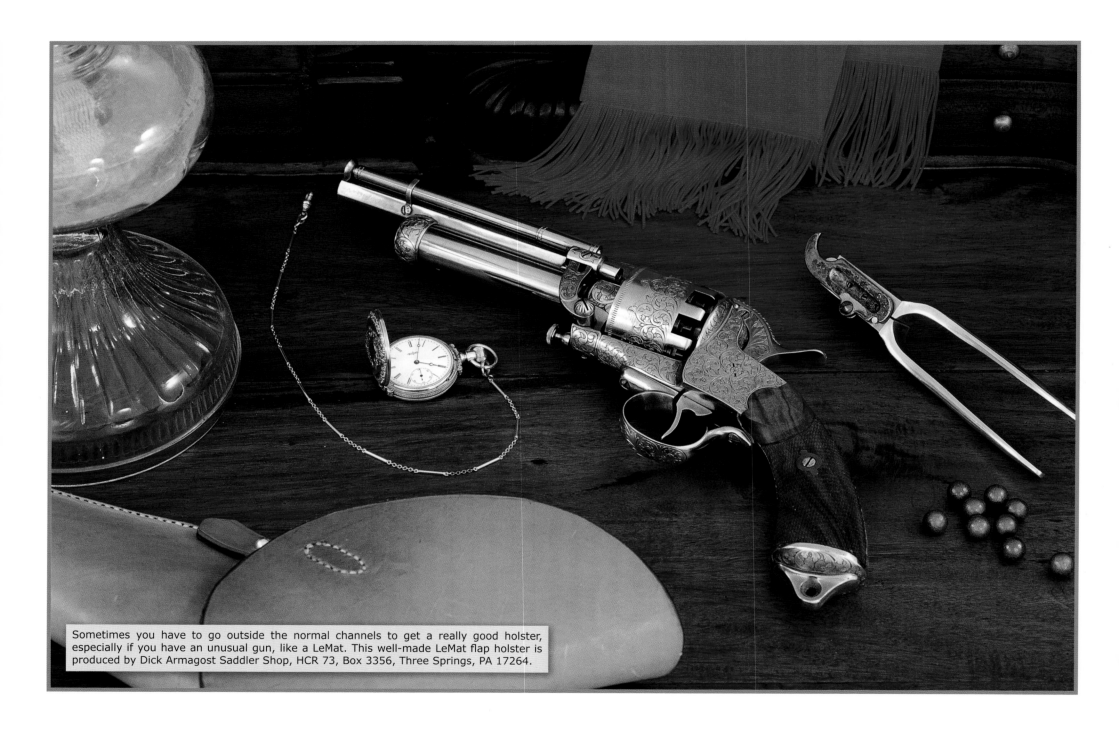

Sometimes you have to go outside the normal channels to get a really good holster, especially if you have an unusual gun, like a LeMat. This well-made LeMat flap holster is produced by Dick Armagost Saddler Shop, HCR 73, Box 3356, Three Springs, PA 17264.

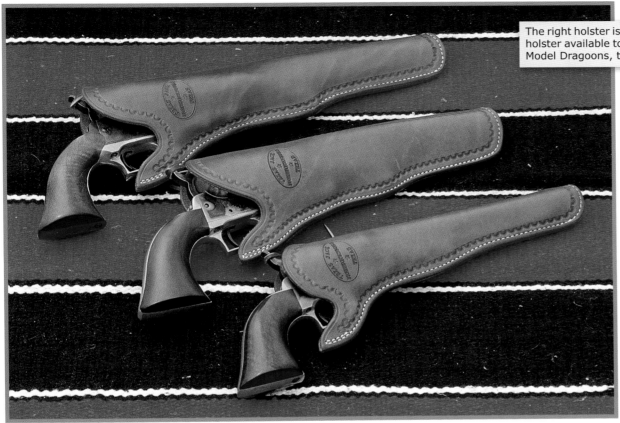

The right holster is a must for percussion pistols. The Slim Jim is a traditional style holster available to fit models ranging from the 1847 Walker, and 1st through 3rd Model Dragoons, to the smaller 1851 Navy. Holsters from Texas Jack's.

Floral pattern COWS holsters for 1860 Army and 1851 Navy models add a touch of class to any shooting iron.

GO WEST YOUNG REENACTOR

The American Frontier has been a fascination with people the world over since the days of the silent movie and actors like the great Tom Mix. More films have been made about the Old West than any other subject, and nearly all of the greatest film legends like John Wayne, Henry Fonda, and James Stewart portrayed cowboys or cavalrymen at one time or another in their careers. And for more than half of a century, television has entertained us with tales of the American West and legendary characters like Wyatt Earp, Wild Bill Hickok, and Bat Masterson. It comes as no surprise then, that Cowboy reenctments are even more popular today than those of the American Revolution and Civil War.

Unlike Civil War reenactments where you shoot blanks, (powder packed over with cream of wheat or soft wads) in most Western events you shoot live ammunition and compete for titles.

Champion shooter and film producer Hunter Scott Anderson, known in Cowboy Action Shooting as "Bounty Hunter," is shown wearing Wah Maker's popular Leadville shirt, black frontier pants, and leather braces.

Known as Cowboy Action Shooting, the events are staged by the Single Action Shooting Society, an international organization created in 1982 to preserve and promote the sport of western shooting. SASS endorses regional matches conducted by affiliated clubs around the country, and annually produces End of Trail, the Nation's oldest and largest Cowboy Action Shooting competition and Wild West festival.

As with Civil War reenactments, authentic clothing and firearms are a must, and SASS contestants must compete with firearms typical of those used in the taming of the Old West. The required use of vintage firearms and authentic clothing make Cowboy Action Shooting one of the most colorful and entertaining shooting sports in the world.

Tom Filback in a Wah Maker Burgundy Cedar Bluff shirt, and olive chaparral pants, takes aim with a Remington .44 caliber revolver.

SKIN THAT SMOKEWAGON!

Whereas percussion revolvers are the only sidearms used in Civil War reenactments (even though a handful of cartridge-firing pistols were being carried during the war), in Cowboy Action Shooting, it is cartridge-firing revolvers that rule the roost, and cap and ball pistols are limited only to the Frontiersman class, which allows percussion pistols manufactured prior to 1896, or reproductions thereof. (This category is also open to Ruger Old Army revolvers without adjustable sights).

Duelist is a traditional shooting style defined by SASS as "a single action revolver fired one-handed and unsupported." It is pure Old West gunfighter shooting, and in the Side Match Plainsman event, shooters carry two percussion revolvers which must be fired one-handed, much in the style of legendary shootist Wild Bill Hickok, who carried a pair of 1851 Navy revolvers.

Tom Filback in black gabardine frock coat, tan saddle embroidered bib shirt and Dodge City striped pants from Wah Maker.

Within the Frontiersman class, the rules are simple, pistols must be .36 caliber or larger, and all clothing, holsters, and accoutrements must be correct for the period. To quote the SASS handbook: "Cowboy Shooting is a combination of historical reenactment and Saturday morning at the matinee. Participants may choose the style of costume they wish to wear, but all clothing must be typical of the late 19th century, a B-Western movie, or Western television series."

Movies and TV open the door for some pretty broad interpretations of Western wear. Not so in Civil War reenactments, where uniforms must conform to the period, regiment, etc., for Union Army, and to the style of Confederate uniforms worn by Southern officers and infantry. However, when it comes to period clothing the Rebs definitely have the most fun with their wardrobes.

The legendary Dennis "China Camp" Ming, six time Cowboy Action Shooting World Champion, is shown with Wah Maker's High Plains bib shirt, teal canvas Frontier pants, harness leather braces, and "Boss of the Plains" hat.

CHAPTER SEVEN: Practical Percussion Pistol Shooting

Veteran Cowboy Action Shooters Hunter Scott Anderson, China Camp, and Tom Filback in action at a Cowboy Action Shooting match in the late 1990s.

Unless you have a friend in the costume rental business, access to a movie studio wardrobe department, or a very talented seamstress in the family, the only way to get authentic period clothing is to buy it.

Getting involved in Civil War reenactments and Cowboy Action Shooting is an exciting counterpart to collecting Colt Blackpowder reproductions and replica revolvers. These events provide an opportunity to experience first hand the era in which such fabled handguns as Colt's 1860 Army and Remington's innovative .44 were used. At the same time, these historic reenactments enrich our knowledge of the past, and the legendary role percussion revolvers played in our country's history, and the settling of the American West.

Dressed to kill... "Bounty Hunter" Scott Anderson wears Wah Maker's Bozeman vest, Lonestar shirt, and black canvas Frontier pants.

Wah Maker's Trail Duster coat gives Hunter Scott Anderson a true Western look. Dennis Ming wears a brown wool herringbone frock coat, Latigo vest and white victorian shirt, and Tom Filback is decked out in a black Wah Maker gabardine frock coat, tan saddle embroidered bib shirt and striped Dodge City pants.

Suitable for either short barrel, medium frame percussion pistols, or metallic cartridge conversion pistols with 5 inch barrels, the COWS old west-style Huckleberry Rig is a fine looking piece of leather work, and a darn quick draw to boot. This was the style holster worn by Val Kilmer in *Tombstone*.

Frontier clothier Allen Wah and champion Cowboy Action shooter Hunter Scott Anderson.

CHAPTER SEVEN: Practical Percussion Pistol Shooting

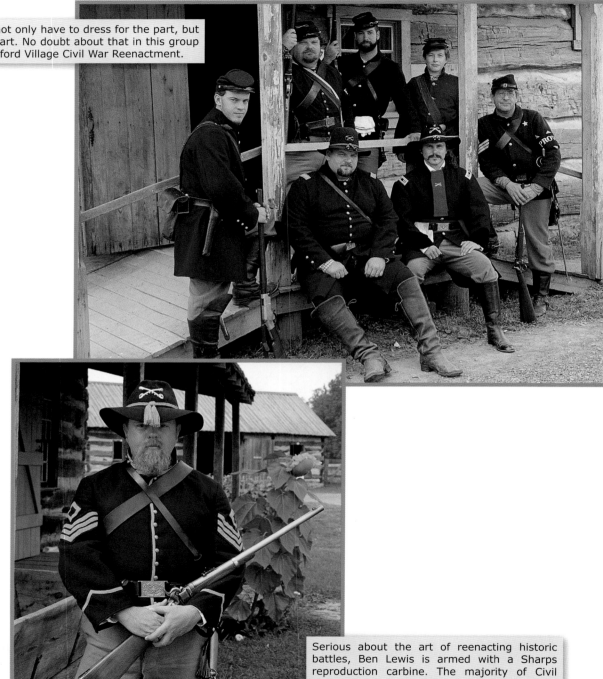

To be believable, you not only have to dress for the part, but you have to look the part. No doubt about that in this group from the 1997 Old Bedford Village Civil War Reenactment.

Back in the late 1990s Chris E. Shockey, a young Civil War enthusiast started playing the roll of Capt. James H. Kidd, Cmdr. Co. E., 6th Michigan Volunteer Cavalry, in reenactments at Gettysburg and throughout Pennsylvania.

Serious about the art of reenacting historic battles, Ben Lewis is armed with a Sharps reproduction carbine. The majority of Civil War reenactors research their uniforms and firearms, as well as the look of the period.

One of the most colorful figures of the Civil War and American West was George Armstrong Custer. Bedford's resident Custer reenactor, Joe Topinka, has played the role for TV documentaries, magazine articles and in Civil War events from Gettysburg to Bedford.

Tombstone outfit from Texas Jack's features Wah Maker stand up collar gambler shirt, tie, paisley vest and black frock coat. Western badge manufactured by Bruce Daly.

Corporal Mark Anderson takes careful aim with his .31 caliber Roots pistol. The small revolvers were sometimes carried by infantry and officers as a back-up pistol. Reproductions of the Roots pistol, rifle and carbine were sold by Dixie Gun Works.

CHAPTER SEVEN: Practical Percussion Pistol Shooting

The author in red COWS narrow bib shirt, black canvas pants from Wah Maker, cavalry boots, paisley Wild Rag, and leather cuffs from Texas Jack's. Custom knife by George Baseke. Gunbelt and Slim Jim holster from COWS.

Cowboy Action Shooter and Civil War reenactor Robert Russell is shown in authentic Western wear from Texas Jack's in Fredericksburg, Texas.

Wah Maker duster gives Bob Russell an even more sinister look. Holster and gunbelt from COWS and Texas Jack's.

That's George Baseke dressed in traditional western garb. Everything but George and his hat are available from Texas Jack's.

Robert Russell in Wah Maker clothing from Texas Jack's. Bob is taking aim with a Uberti 1847 Walker. Hat from Texas Jack's in Fredericksburg, TX.

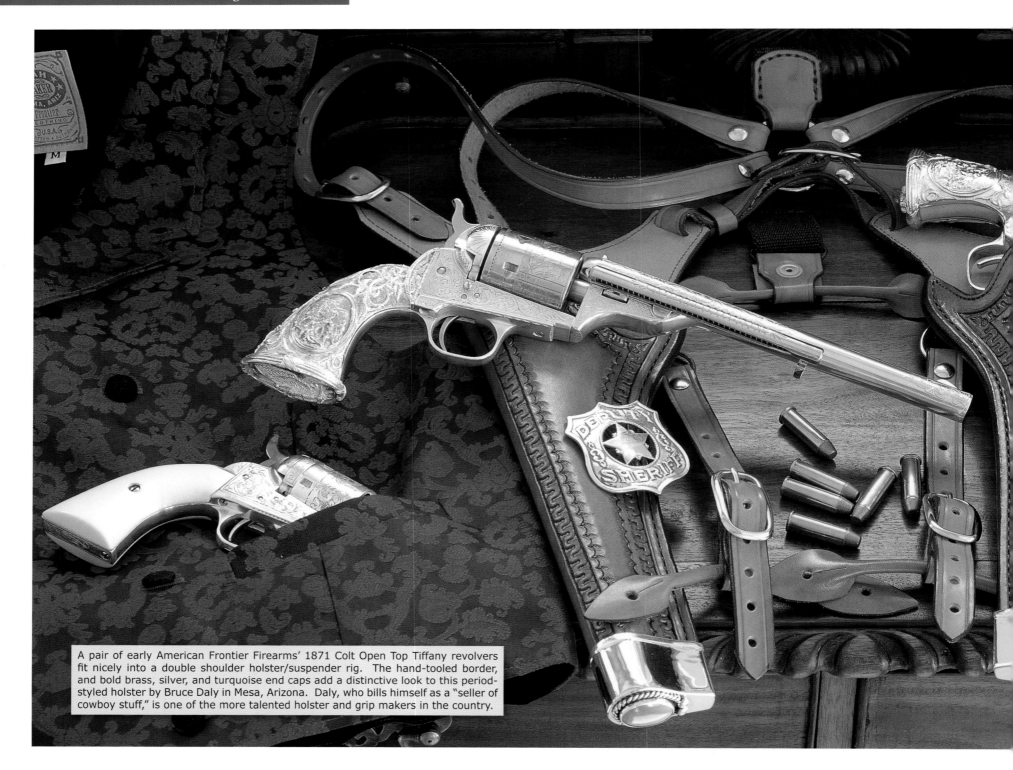

A pair of early American Frontier Firearms' 1871 Colt Open Top Tiffany revolvers fit nicely into a double shoulder holster/suspender rig. The hand-tooled border, and bold brass, silver, and turquoise end caps add a distinctive look to this period-styled holster by Bruce Daly in Mesa, Arizona. Daly, who bills himself as a "seller of cowboy stuff," is one of the more talented holster and grip makers in the country.

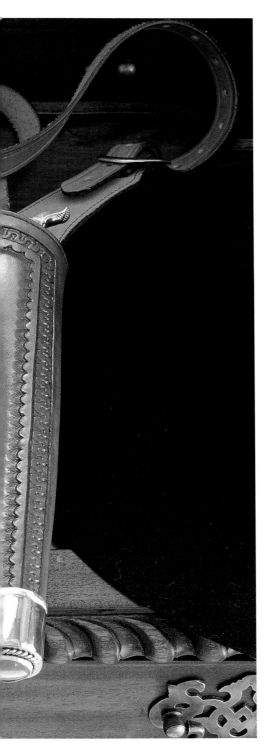

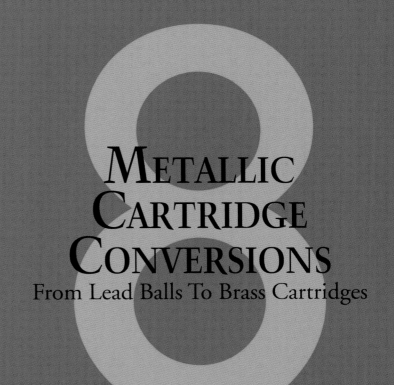

METALLIC CARTRIDGE CONVERSIONS
From Lead Balls To Brass Cartridges

The story of the Old West, at least the Old West most of us have come to know through movies and television, was populated with good guys and bad guys who all seemed to have one thing in common, the same type of gun.

For the longest time we believed that the Colt Single Action Army, the good old Peacemaker, was "The Gun that Won the West." Few will argue the SAA's role in the history of the American Frontier, but from the end of the Civil War in 1865 to as late as 1878, most of the "winning" was done with Colt percussion pistols and the first American-made cartridge-firing revolvers, the Richards, and Richards-Mason Colt conversions.

With the end of Smith & Wesson's domination of the metallic cartridge market in America, in 1869, (this through their patent on bored-through revolver chambers), Colt produced its first factory cartridge conversion model, the 1871-72 Open Top. This was followed in 1873 with the introduction of the revolutionary Single Action Army, which was to become the most successful hand gun of the 19th century. However, concurrent with the Open Top, conversions of the 1851 Navy, 1860 Army, and Colt Pocket Pistols helped lead the way Westward and as more affordable alternatives to the new Peacemaker, the cartridge conversion models would see use well into the 1880s.

THE FIRST CARTRIDGE PISTOLS

Although the Colt and Remington percussion pistols were regarded as state of the art in the United States, cartridge-firing revolvers had been in use throughout Europe since the 1840s, and were carried by both Union and Confederate troops during the Civil War.

The type of bullet developed in Europe was known as the pinfire, which literally meant that the firing pin was in each individual bullet. The hammer struck the pin, which in turn struck a percussion cap inside the cartridge, igniting the powder.

CHAPTER EIGHT: Metallic Cartridge Conversions

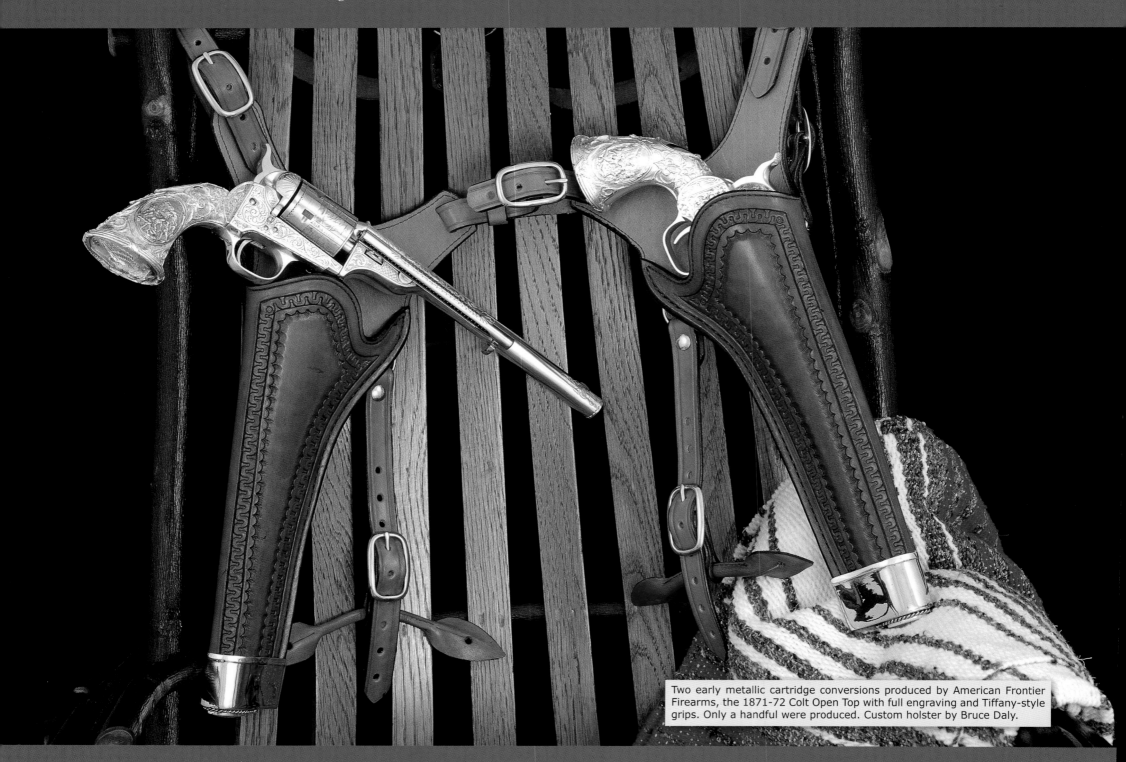

Two early metallic cartridge conversions produced by American Frontier Firearms, the 1871-72 Colt Open Top with full engraving and Tiffany-style grips. Only a handful were produced. Custom holster by Bruce Daly.

Pictured is an early American Frontier Firearms 1861 Navy model with Richards Type I conversion in .38 caliber. This was a hybrid design by Anderson as there were no Richards I conversions originally done on 1861 Navy models.

Anderson built Remington conversions with one-piece cylinders. The guns were offered in .44 Russian and .38 special. Like many early conversions by Dave Anderson, these are hard to come by today.

The author lets loose with an American Frontier Firearms 1851 Navy Richards conversion in .44 Russian. The author's custom Colt 1861 Navy is holstered in a COWS Huckleberry shoulder rig.

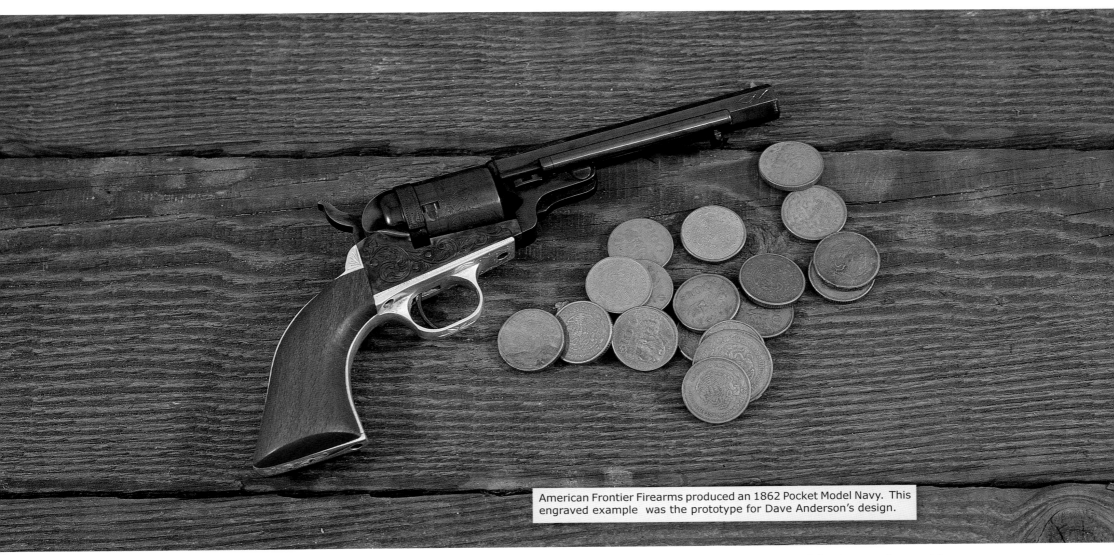

American Frontier Firearms produced an 1862 Pocket Model Navy. This engraved example was the prototype for Dave Anderson's design.

In 1862 President Abraham Lincoln commissioned Marcellus Hartly, a partner in the New York firearms importing firm of Schuyler, Hartley & Graham, to supply the Union Army with French Lefaucheux (*lue-foe-sho*) pistols and pinfire ammunition. The Lefaucheux were the fourth most commonly used revolver in the American Civil War, surpassed only by the Colt, Remington, and Starr percussion revolvers. The Union Army ordered 1,000,000 pinfire cartridges, the largest single order for metallic cartridges in the Civil War. (The Smith & Wesson patent in the U.S. prevented Colt and Remington from developing breech-loading, cartridge-firing models, but it did not prevent the importing of foreign-made cartridge-firing revolvers. The Confederate Army also used Lefaucheux pistols and a small number of LeMats produced in France were chambered for pinfire cartridges).

It was Richards and Mason that really put the lead into Colt's legendary percussion models, with their patented conversion to .44 Colt centerfire cartridges. Between 1871 and 1878 more than 10,000 Colts were converted to .44 caliber and roughly another 6,000 to .38 caliber rimfire and centerfire

CHAPTER EIGHT: *Metallic Cartridge Conversions*

An original early Richards conversion without ejector (top) alongside an American Frontier Firearms 1860 Army conversion.

rounds, the latter, principally 1851 Navy and 1861 Navy models that had originally been .36 caliber cap and ball revolvers.

The original Richards conversion used a breech ring behind a modified cylinder, (the back portion of which was cut off and the chambers bored through) with the firing pin held within the breech ring itself. The later Richards-Mason version had a hammer-mounted firing pin and used newly-manufactured barrels which differed from earlier percussion models.

Many of the conversions were done by the Colt factory, while others were produced by independent gunsmiths who purchased old Colts and rebuilt

Taken in 1997, gunmaker Dave Anderson is shown working on the author's custom 1861 Colt Navy with .38 caliber Richards Type I conversion. Still designing guns today, Anderson is creating special Colt SAA-style revolvers tailored to the needs of Mounted Shooting compeition.

CHAPTER EIGHT: Metallic Cartridge Conversions

One of the neatest cartridge conversions available is done by R. L. Millington on the Starr single or double action revolvers manufactured by Pietta. Using the same technique as those built for metallic cartridges in the 1870s, the Millington Starr conversion is not only a rugged gun but very accurate out to 50 feet grouping five rounds within an inch from center to center. The Starr's unique cylinder design did away with the conventional center arbor (around which both Colt and Remington cylinders revolved), and instead the long ratchet shaft seated into the breech at the rear and locked into the front of the frame with the conical bolt extending from the cylinder.

them to fire the new .38 and .44 cartridges. This was a booming business in the early 1870s, as the settling of the American West commenced.

Original Colt conversions are readily available as are black powder cartridges for these guns produced by Ten-X, Goex, and Black Hills, as well as for the handcrafted reproductions manufactured by such notable gunsmiths Kenny Howell in Beloit, Wisconsin, and production versions built in Italy by Uberti.

The latest models to come to market use modern smokeless powder ammunition in standard calibers of .38, .44 Colt, .44 Russian and .45 Colt. Popular among Cowboy Action Shooters, and just about anyone who has seen one of these striking pistols in a movie or TV Western, the modern production Colt cartridge conversions were pioneered in the late 1990s by gunsmith Dave Anderson and his company American Frontier Firearms.

A former machinist, Anderson got into gunsmithing as a second career in the late 1970s, and eventually purchased a gun shop in Lakewood, California. It was there that he began to build custom pistols and rifles, and specialized equipment for the Los Angeles Police Department's SWAT Team.

"I always liked Western guns the best," says Anderson, "and when the Single Action Shooting Society was formed I got in on the ground floor of that, building custom speed guns and competition Western pistols."

Anderson researched the Richards and Mason patents, and decided to follow their general design in the production of Colt and Remington conversions for use in Cowboy Action Shooting. "Over a long period of time I started converting Italian guns like other people have done, and trying to figure out how to build them to fire conventional smokeless powder cartridges, basically off the shelf ammo, rather than custom black powder loads. That was really what we felt the market demanded, if conversion guns were to become practical shooters."

Anderson's early Colt and Remington conversions were featured in a number of movies, including *The Quick and the Dead*, and in television shows like the long-running series *Dr. Quinn, Medicine Woman*.

Producing authentic-looking period handguns was something of a crusade for Anderson when he started American Frontier Firearms. From the original 1851 Navy and 1860 Army conversions, he expanded into a variety of Colt Pocket Pistols, (which comprised the greatest number of original conversions done in the 1870s), and Remingtons. It was from many of his original designs that the Italian reproductions we see today evolved over the last decade.

Anderson also designed his own version of the Richards breech ring with a floating firing pin. Like Colt Blackpowder Arms Co., American Frontier Firearms did all of the final fitting, bluing, and assembly of the guns in the United States. "Each was hand built," explains Anderson, "and in some cases custom tailored to the customer." An excellent example is the author's 1861 Navy, a hybrid Richards design which Anderson built as a short-barreled revolver chambered in .38 caliber with a nickel finish, nitrite blued screws and wedge, and white Micarta stocks.

The most popular model that Anderson built was the 1851 Richards-Mason Navy conversion, and the 1871-72 Open Top. "The 1860 Army and 1861 Navy were the next most popular, but it's the original 1851 Navy that seems to get the most orders." While that has changed over the years to favor the 1860 Army conversion, the Navy models are still in high demand for shooters who like a little less lead in their guns.

The author's two AFF conversions, 1861 Navy in .38 caliber and 1851 Navy in .44 Russian, have both proven to be excellent shooters over the last decade and of a quality of fit and finish that equaled the Colt Blackpowder line. Since the days when Anderson was the only mass producer of conversions, along with handcrafted guns by Kenny Howell and R. L. Millington, the Italian copies have dominated the market with a broad variety of Colt and Remington conversions sold through Cimarron F.A. Co., Taylor's & Co. and Dixie Gun Works, among others.

A Starr is Born

One of the very best conversions from percussion to metallic cartridge was the Starr top break single action revolver. The .44 caliber Starr was second only to the Colt 1860 Army and Remington Army revolvers among U.S. troops. The Federal Government purchased 47,952 Starr double and single action revolvers from the New York arms maker during the Civil War.

Pictured are two custom-built conversions done by Dave Anderson in the late 1990s, an 1862 Pocket Police Richards-Mason Conversion, and an 1848 Baby Dragoon.

With the expiration of the Colt's patent in 1857, Eben T. Starr was one of many gunmakers anxious to start manufacturing revolvers. Most everyone tried to copy Colt, except for Remington, which took a different approach building a solid frame revolver with a topstrap, and Eben Starr, who was literally years ahead of everyone with a double action top break revolver. Starr also designed and patented a breech loading carbine similar to the Sharps, but actually better designed and later converted to fire metallic cartridges.

The first Starr double action revolvers were chambered in .36 caliber. By 1862 demand from the War Department brought about the addition of a .44 caliber version, which amounted to 16,100 guns by May of 1863. A single action version with a longer 8 inch barrel was also requested (the double action models had 6 inch barrels) and this became the most prolific of the Civil War era Starr arms, with production of the 1863 single action model reaching 25,000 by the end of 1864.

The advantage of the Starr as both a percussion arm and later as a cartridge conversion was the top break design. Although Eben Starr didn't patent the idea (Sam Colt had actually proposed such a design among his 1850 Dragoon patents) Starr made improvements to the concept by mortising the top strap to fit over the standing breech, thus giving his guns incredible strength. Starr revolvers were built to withstand the punishment of heavy use, yet by simply unscrewing the large knurled cross bolt that passed through the breech, the barrel and topstrap, which were hinged at the bottom of the frame, to pivot down allowing an empty cylinder to be replaced in a matter of seconds. The gun could also be loaded conventionally using the rammer and plunger. Only a Remington was faster to reload.

As a conversion pistol the Starr was a natural. Most were converted by individual gunsmiths (Starr went out of business in 1867) while others are reputed to have been converted in Belgium for sale to Germany. Both single and double action models were converted to fire cartridges and there were two ways in which the guns were modified, which appears to be a constant, either as a 5-shot, .45 caliber Benet primed centerfire cartridge revolver (mostly double action models) or a 6-shot, .44 Colt conversion, which comprised the majority of guns. The example pictured was done by R. L. Millington on a Pietta Starr single action model copying the design from an original Starr conversion. This style of conversion is the predominant type.

The Starr's unique cylinder design did away with the conventional center arbor (around which both Colt and Remington cylinders revolved), and instead the long ratchet shaft seated into the breech at the rear and locked into the front of the frame with a conical bolt extending from the cylinder. Because of this design Starr revolvers were less apt to foul since they did not have to rotate around an arbor, and as a conversion, the cylinders offered the same advantage. Explains Millington, "A new bored through cylinder had to be made with six bolt stops (the percussion cylinders had 12 stops, one for each chamber and a safety stop in between). A new cylinder ratchet was either machined or taken from the percussion cylinder and attached. The percussion cylinder ratchet passed through the bored out breechring, which contained a floating firing pin like a Richards Type 1 Colt conversion. A relief cut was then made in the front of the hammer to strike the firing pin. The hammer sight notch was untouched. They were very simple conversions because there is was no cutout for loading and no ejector housing. You simply broke open the gun and removed the cylinder to load it."

As a cartridge gun the Starr is exemplary with all the durability of a solid frame revolver and the loading ease of a Remington.

Drop-In Conversions

Over the last decade the advent of "drop-in" conversions for Remington, Colt, and Ruger Old Army black powder revolvers has opened new doors for shooters, allowing them to literally have their cake and eat it too. One of the pioneers of this concept was Minneapolis engineer Walt Kirst who began offering the "Kirst Konverter" for Remington black powder revolvers in the late 1990s. The original Kirst system was an adaptation and improvement of several designs patented in the late 1850s; principally the C. C. Tevis patent (May 17, 1856), J. Adams patent (July 12, 1861), and W. Tranter patent (July 20, 1865)[1]. All three designs were of British origin, utilizing a two-piece cylinder. Each had ratchets on the back plate so that it rotated the cylinder, and the hammer was able to strike the cartridge rim or primer by protruding through notches cut into the back plate. Kirst reasoned that Colt's Richards Type I conversion, patented in July 1871, was a better design with its fixed breech ring and integral, rebounding firing pin, so he combined that with the British designs to create the first

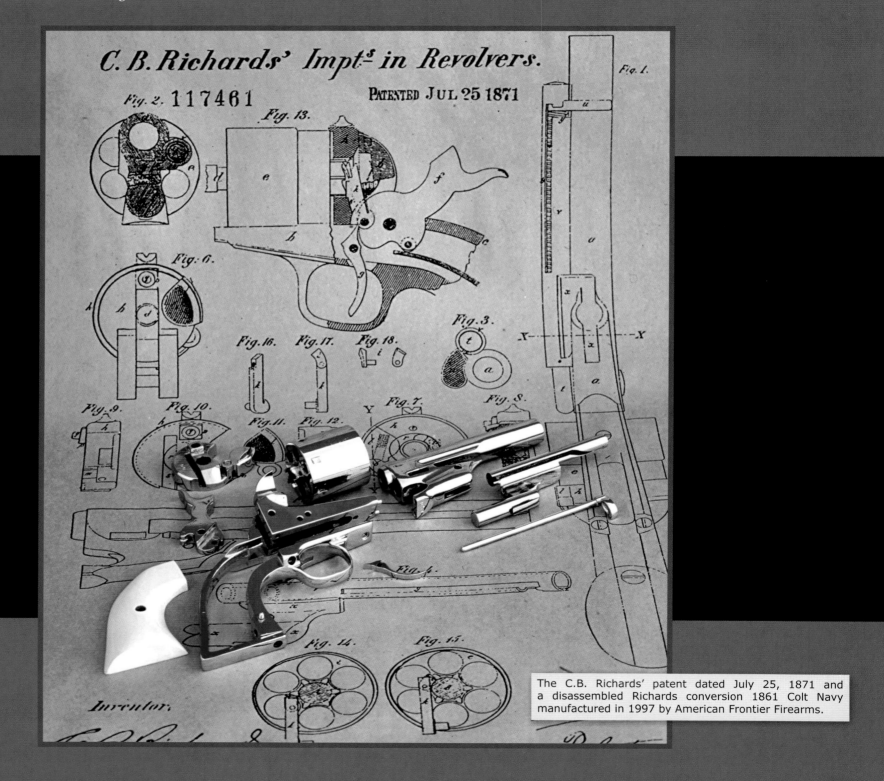

The C.B. Richards' patent dated July 25, 1871 and a disassembled Richards conversion 1861 Colt Navy manufactured in 1997 by American Frontier Firearms.

Kirst conversion cylinder. The design integrated the best aspects of both the Adams & Tranter two-piece cylinders with the Richards Type I. To keep the back plate (breech ring) from rotating, Kirst simply designed it with a flat bottom, which immediately locks and registers the breech ring in place with the frame as soon as the two-piece cylinder is dropped in. The cylinder ratchet simply passes through the breech ring allowing it to rotate as normal. You have to wonder how this idea escaped Remington designers over 150 years ago when it really would have made a difference.

The Kirst conversion cylinder for Remingtons was an immediate hit in 1999, but Walt had another idea. In 1868, Remington struck a deal with Smith & Wesson to produce a cartridge conversion model of the Remington New Model Army; the Ilion, New York gunmaker agreed to pay S&W a licensing fee of $1.00 per gun. The Remington revolvers were converted with the addition of a narrow, fixed breech ring, new cylinder chambered for the .46 caliber rimfire metallic cartridge, and a channeled recoil shield allowing for the loading and unloading of the cylinder without removing it from the gun. This was the inspiration for Kirst's second model. His new Remington and Colt conversion cylinders have a channeled breech ring. The only problem is that the Remington Army and Colt 1851 Navy cap and ball revolvers do not have channeled recoil shields! Not to worry. The Kirst System comes complete with cylinder, channeled breech ring, and for Remington models, a new center pin for the cylinder with an attached ejector. All you need to do is channel the recoil shield. This is a pretty heroic suggestion to anyone who has never tried it, but Kirst has made it relatively simple, even for a novice. All that is required are simple tools: a couple of high quality round files, a 1/4 inch wide flat file, 3M "00" finishing pads, "0000" super fine steel wool, a Dremel tool, and a bottle of Birchwood Casey Super Blue cold blue. The step-by-step instructions supplied by Kirst are easy to follow and include a template that provides the exact positioning and depth of the recoil shield channel. With the proper tools, you can start the project in the morning and be on the shooting range before the sun sets. And there is a feeling of accomplishment when you're finished that you can't get with a checkbook.

Not long after Kirst's cylinders hit the market, world renowned gunmaker Kenny Howell (who builds all of the movie guns for Tom Selleck's westerns), teamed up with Taylor's & Co. to produce a drop in two-piece cylinder with six firing pins on the top cap. This design, closer to the original British patents of the 19th century, provided an even more efficient way to switch out loaded cylinders for Cowboy Action Shooting with a Remington .44 caliber black powder revolver. Howell and Kirst have both added the Ruger Old Army to their list of drop-in conversion cylinders, as well as adding limited edition semi-custom conversions in the past year for the 1851 and 1861 Navy.

This brings the history of Colt black powder and Remington repeaters full circle, from the early days of Sam Colt in Paterson, New Jersey, to the final years of the percussion revolver in Hartford, Connecticut and Ilion, New York, to the advent of metallic cartridge conversions in the 1870s and the guns that really did "Win the West."

CHAPTER EIGHT: *Metallic Cartridge Conversions*

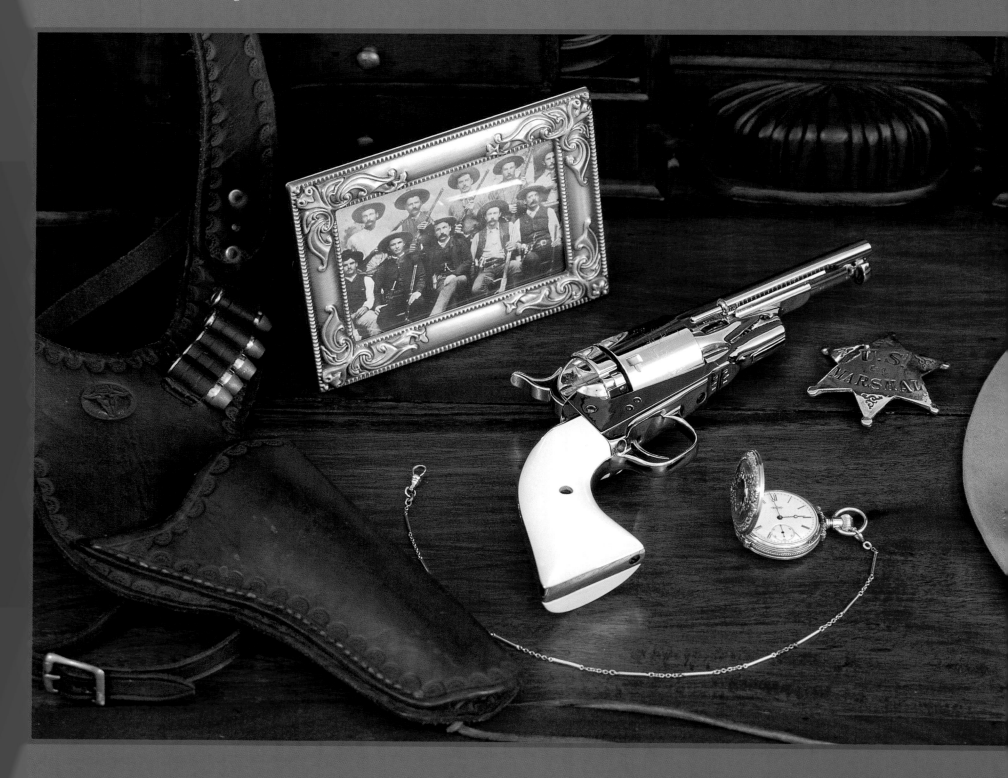

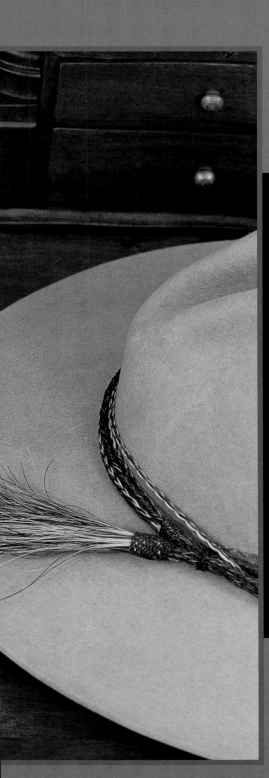

The author's custom-built Colt 1861 Navy with 4-3/4 inch barrel, nickel finish and nitrite blued screws and wedge. This was a fully custom-built Richards Type I built by Dave Anderson in 1997.

206 CHAPTER EIGHT: *Metallic Cartridge Conversions*

Anderson was the first to make an Open Top model in the late 1990s.

Black Powder Revolvers - Reproductions & Replicas

CHAPTER EIGHT: *Metallic Cartridge Conversions*

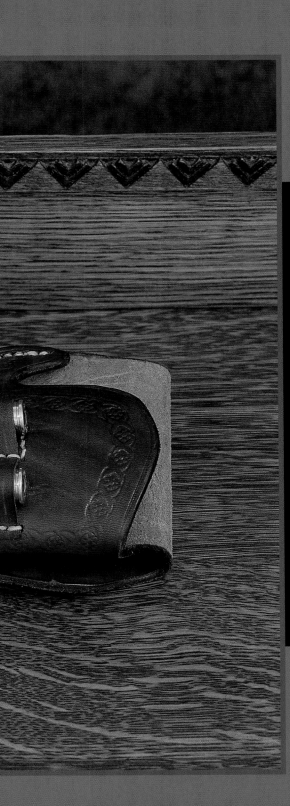

Another model built for the author by Dave Anderson, this was an 1851 Navy conversion in a combined Richards Type 1 and Richards-Mason style and chambered in .44 Russian. The gun utilized the Richards-style breech ring with internal firing pin and rear sight, and Richards-Mason ejector. Holster by Dick Armagost, ammunition by Black Hills.

CHAPTER EIGHT: Metallic Cartridge Conversions

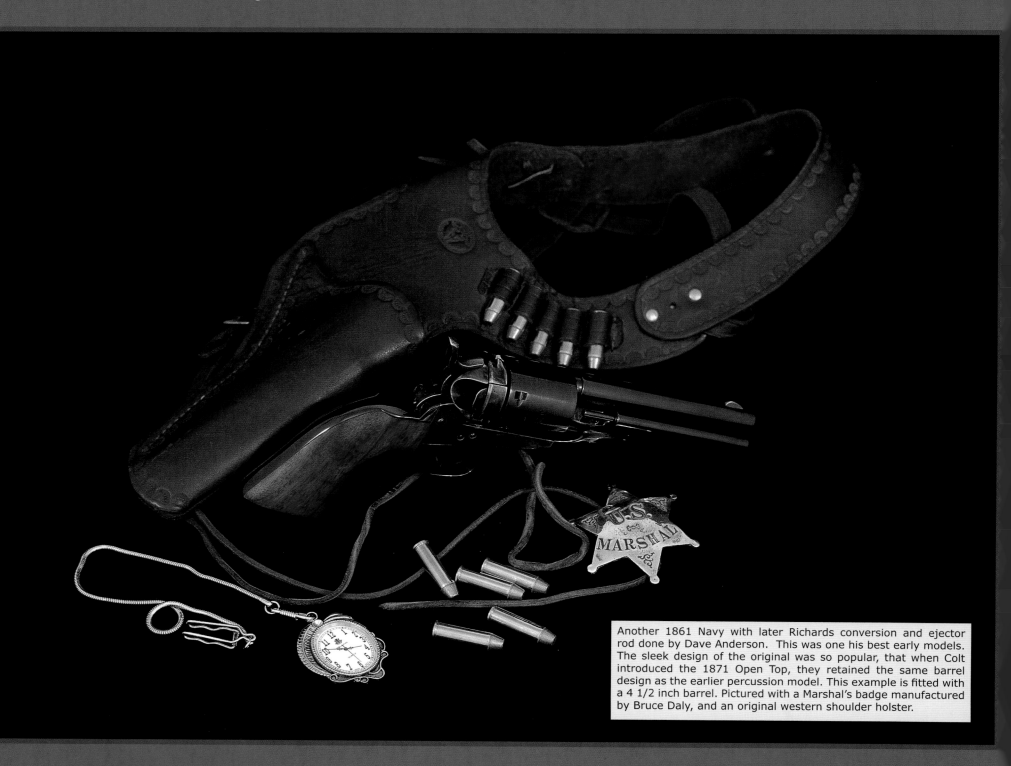

Another 1861 Navy with later Richards conversion and ejector rod done by Dave Anderson. This was one his best early models. The sleek design of the original was so popular, that when Colt introduced the 1871 Open Top, they retained the same barrel design as the earlier percussion model. This example is fitted with a 4 1/2 inch barrel. Pictured with a Marshal's badge manufactured by Bruce Daly, and an original western shoulder holster.

Innovative from the start, Anderson offered the Richards 1860 Army conversion with either 5 1/2 or 7 1/2 inch barrels and in two barrel sets. Most of the tooling used to make guns for American Frontier Firearms formed the basis for the Italian conversions built today.

CHAPTER EIGHT: Metallic Cartridge Conversions

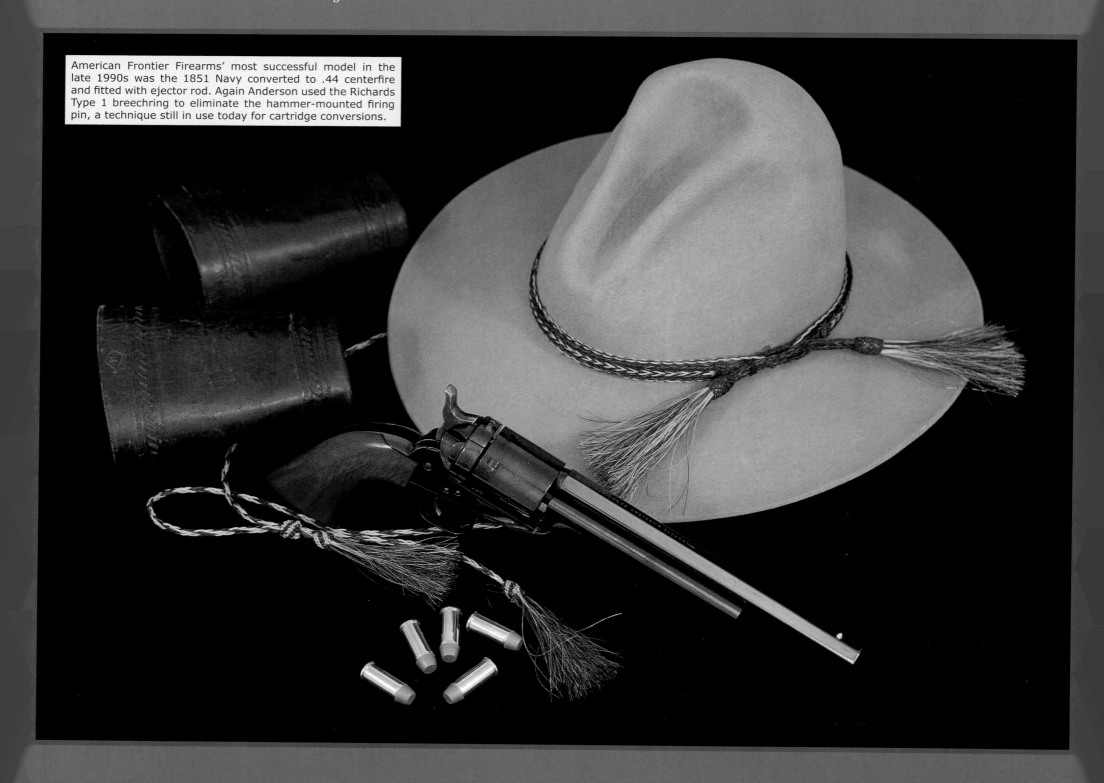

American Frontier Firearms' most successful model in the late 1990s was the 1851 Navy converted to .44 centerfire and fitted with ejector rod. Again Anderson used the Richards Type 1 breechring to eliminate the hammer-mounted firing pin, a technique still in use today for cartridge conversions.

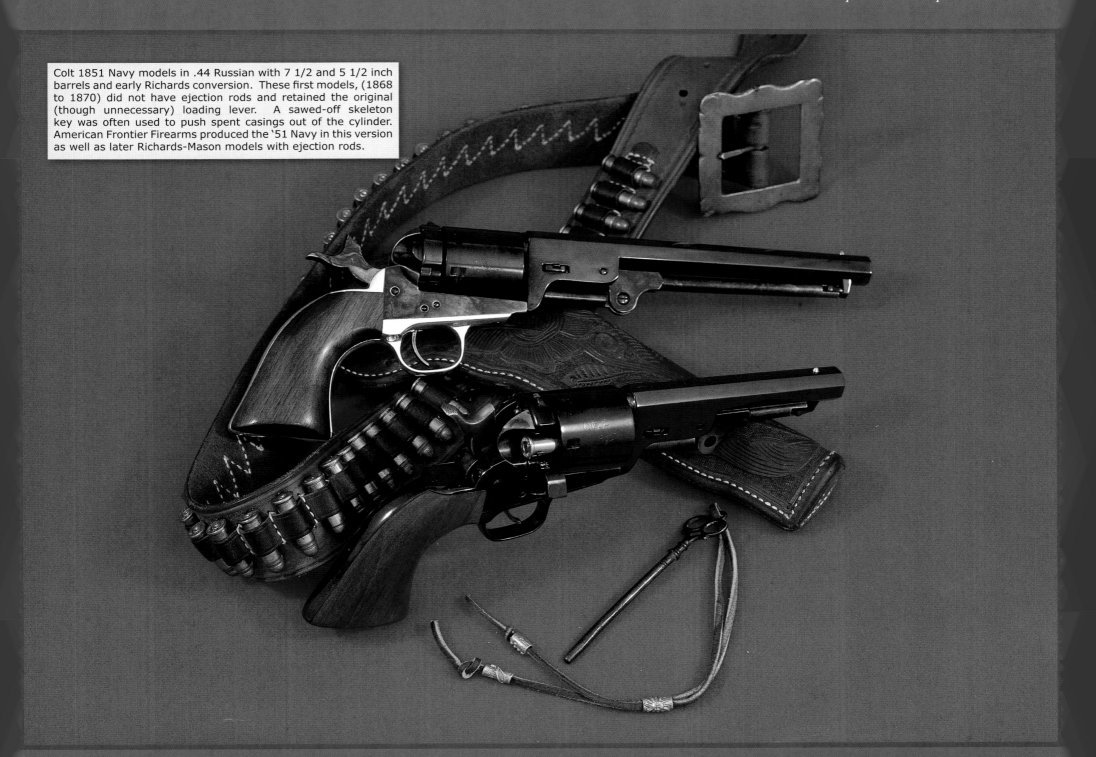

Colt 1851 Navy models in .44 Russian with 7 1/2 and 5 1/2 inch barrels and early Richards conversion. These first models, (1868 to 1870) did not have ejection rods and retained the original (though unnecessary) loading lever. A sawed-off skeleton key was often used to push spent casings out of the cylinder. American Frontier Firearms produced the '51 Navy in this version as well as later Richards-Mason models with ejection rods.

CHAPTER EIGHT: Metallic Cartridge Conversions

Among the most common and most popular Colt conversions of the 1870s were the Richards-Mason factory conversions of the 1860 Army (top) and 1862 Police model. Both guns pictured were handcrafted and engraved by R. L. Millington of ArmSport LLC in Platteville, Colorado using 3rd Generation Colt revolvers. While there are commercially-produced 1860 Army conversions on the market built in Italy, the Pocket models are available only from Millington.

CHAPTER EIGHT: *Metallic Cartridge Conversions*

R. L. Millington was the first to produce an authentic Richards Type I conversion of the 1860 Army. Though fewer than 50 guns have been built each is handcrafted and most have been engraved either by Millington or by Conrad Anderson of Rocktree Ranch (such as the nickel and gold example) in Kingston, Idaho.

Black Powder Revolvers - Reproductions & Replicas 217

Cased Richards Type I conversions in .44 Colt were produced for the author (nickel gun) and two private collectors (gold and nickel gun) and are among the most limited and expensive contemporary Colt conversions to date.

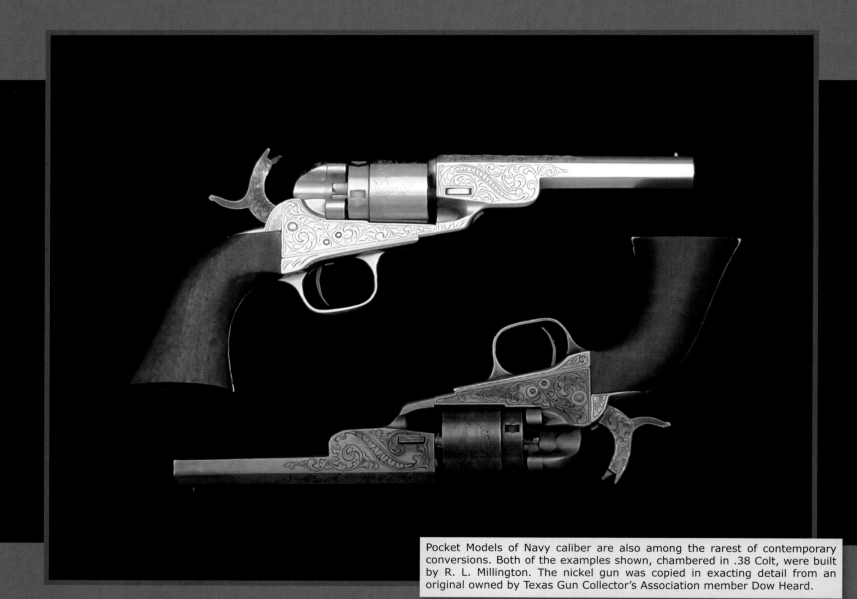

Pocket Models of Navy caliber are also among the rarest of contemporary conversions. Both of the examples shown, chambered in .38 Colt, were built by R. L. Millington. The nickel gun was copied in exacting detail from an original owned by Texas Gun Collector's Association member Dow Heard.

The most elaborate contemporary Richards Type I conversion ever built, this Tiffany model was handcrafted by R. L. Millington and hand engraved and Tiffany gripped by John J. Adams Sr. for the book cover of *Colt Single Action – From Patersons to Peacemakers*. The custom holster was made by Jim Lockwood of Legends in Leather. Following the book's debut at the Colorado Gun Collector's Show in May 2007, the gun was lost by US Airways which mistakenly shipped it to Africa, instead of Pennsylvania, where it disappeared somewhere in Ethiopia.

CHAPTER EIGHT: Metallic Cartridge Conversions

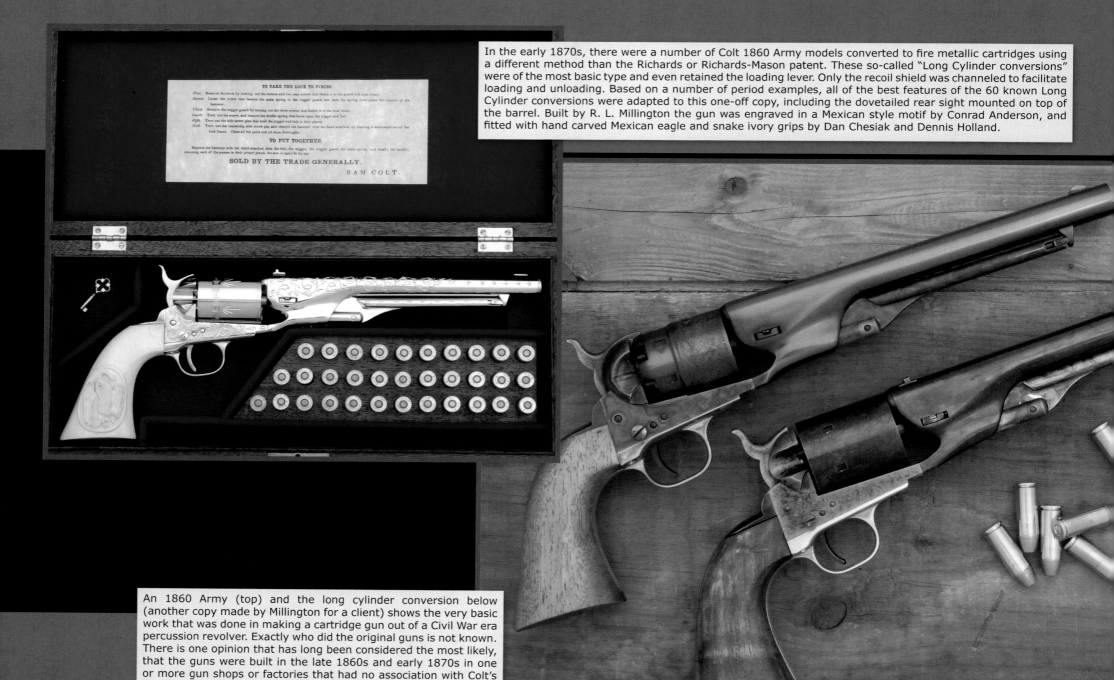

In the early 1870s, there were a number of Colt 1860 Army models converted to fire metallic cartridges using a different method than the Richards or Richards-Mason patent. These so-called "Long Cylinder conversions" were of the most basic type and even retained the loading lever. Only the recoil shield was channeled to facilitate loading and unloading. Based on a number of period examples, all of the best features of the 60 known Long Cylinder conversions were adapted to this one-off copy, including the dovetailed rear sight mounted on top of the barrel. Built by R. L. Millington the gun was engraved in a Mexican style motif by Conrad Anderson, and fitted with hand carved Mexican eagle and snake ivory grips by Dan Chesiak and Dennis Holland.

An 1860 Army (top) and the long cylinder conversion below (another copy made by Millington for a client) shows the very basic work that was done in making a cartridge gun out of a Civil War era percussion revolver. Exactly who did the original guns is not known. There is one opinion that has long been considered the most likely, that the guns were built in the late 1860s and early 1870s in one or more gun shops or factories that had no association with Colt's (such as B. Kittredge & Co. or Schuyler, Hartley & Graham), or that they were hand built in Mexico from surplus 1860 Army parts. Millington has made only a few Long Cylinder conversions.

Among the quality production models brought to market in the last decade is the Uberti built 1871-72 Colt Open Top, which is essentially a finer version of the Long Cylinder design as produced by Colt's in .44 Henry caliber. The Italian copies, like this handsomely engraved model from Cimarron F.A. Co., in Fredericksburg, Texas, are affordable and well built copies of the originals.

The Long Cylinder conversions work efficiently and are easy enough to load and reload. It was a simple but functional design, perhaps unworthy of Colt's in the 1870s, but nonetheless a good cartridge gun.

CHAPTER EIGHT: Metallic Cartridge Conversions

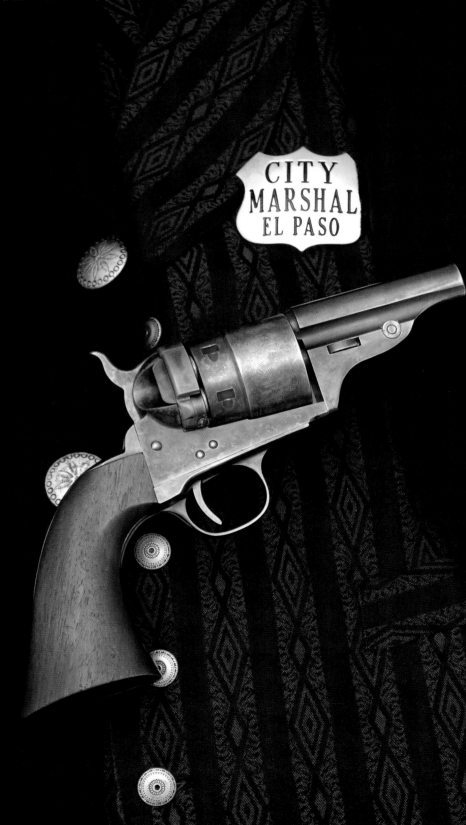

The line that divided lawmen from the lawless was often indistinguishable on the Western Frontier and City Marshal Dallas Stoudenmire walked that line until he was murdered on the streets of El Paso in 1882. Using a Uberti Richards-Mason 1860 Army conversion, R. L. Millington made a duplicate of Studenmire's cut down .44 which he carried in a leather lined trouser pocket. Stoudenmire was a former Texas Ranger and also the U.S. Deputy Marshal for the Southern District of Texas. (Stoudenmire El Paso City Marshal badge by Starpacker Badges, Nashua, New Hampshire)

Cimarron F.A. Co. now has a Richards Type II 1860 Army cartridge conversion chambered in .44 Colt (also in .45 Colt). Built by Uberti, .44 Colt was the original chambering offered with the guns in the 1870s. The Richards Type II was an interim model before the Richards-Mason conversions and utilized a new Mason-designed breechplate and hammer-mounted firing pin, though it still retained the cartridge ejector from the Type I. This is a very elegant looking gun and quite accurate at 50 feet.

CHAPTER EIGHT: Metallic Cartridge Conversions

The 1851 Navy was one of the most prolific of Richards-Mason cartridge conversion guns in the 1870s. This handcrafted reproduction in .38 Colt caliber is manufactured by Kenny Howell and R&D Gun Shop. (Holster and belt by Greg Doring/Ned Buckshots Wild West. Buckle by Tim and Jamie Parsley/Civil War Belt Buckles.)

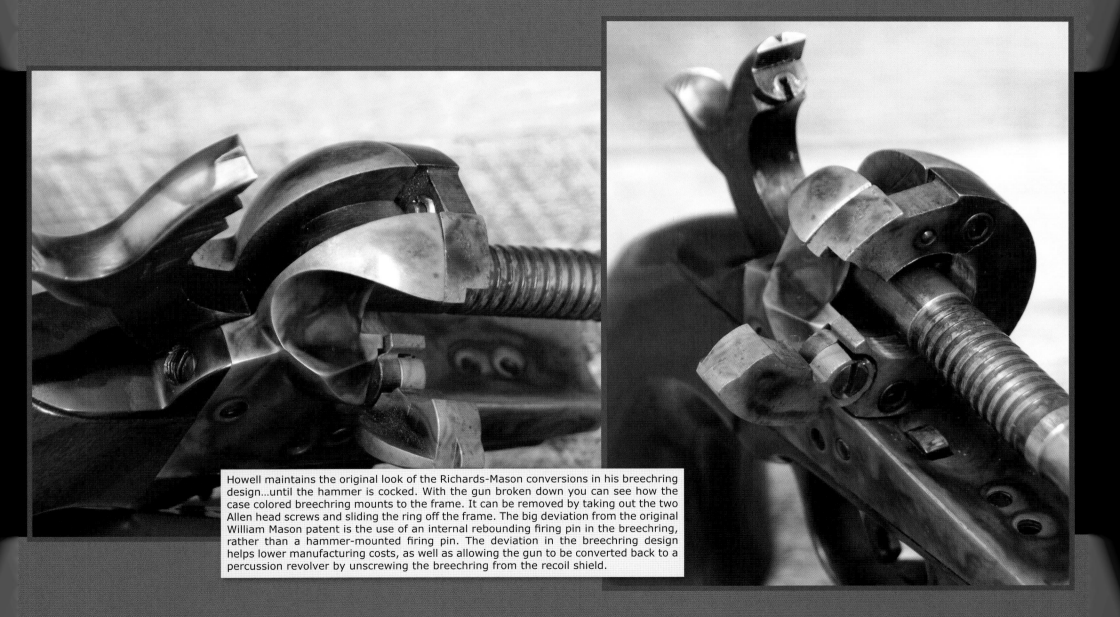

Howell maintains the original look of the Richards-Mason conversions in his breechring design...until the hammer is cocked. With the gun broken down you can see how the case colored breechring mounts to the frame. It can be removed by taking out the two Allen head screws and sliding the ring off the frame. The big deviation from the original William Mason patent is the use of an internal rebounding firing pin in the breechring, rather than a hammer-mounted firing pin. The deviation in the breechring design helps lower manufacturing costs, as well as allowing the gun to be converted back to a percussion revolver by unscrewing the breechring from the recoil shield.

Call it "reverse engineering" from the man who started the drop-in conversion business. The latest model from Walt Kirst is an 1861 Colt Navy conversion that utilizes a firing pin on the hammer, instead of a breechring with an internal floating firing pin like the rest of Kirst's drop-in conversions. "It has a more authentic look when you cock it," says Kirst, "and the firing pin can be unscrewed, the threaded hole filled with the supplied Allen head screw plug, and the gun switched back to a black powder percussion cylinder." The Kirst '61 Navy also features the elegant Richards Type II ejector housing design.

The latest in cartridge conversions cylinder from R&D and Taylor's & Co. encompass two extremes – a drop-in cylinder for the Colt Walker, chambered in .45 Long Colt, and a drop-in for the Colt Baby Dragoon, chambered in .32 S&W. Walt Kirst also has a new addition to his line of conversion cylinders, a .32 S&W for the steel frame Remington pocket pistol. Also pictured is a full c.1870's Richards-Mason conversion of a Colt Walker done for the author by R.L. Millington of ArmSport LLC. (Walker holster by Jim Barnard, Trailrider products)

The Walker drop-in cylinder is easy to load but the drop-in part is a bit more complicated. The gun must be disassembled in order to place the loaded cylinder and breechring on the arbor. If your Walker has a wedge that is easily pushed out the process goes smoothly. Always leave one chamber empty.

This is the original big bang theory. Loaded with Ten-X .45 Colt, 200 gr. RNFP, the Walker packed an accurate wallop placing 5 rounds under an inch at 25 feet.

Sam Colt favored small pistols and after the Walker and 1st Model Dragoon, Colt's introduced the demur Baby Dragoon pocket pistol chambered in .31 caliber. Taylor's & Co. figured what was good on a large scale should work as well on a small one, and had R&D develop a .32 S&W drop-in cylinder for the little Dragoon. At close range this is one nifty little six-shooter. Note the loaded chambers indicated by the brass showing.

Handcrafted by R.L. Millington, this 3rd Generation Walker Colt conversion in .45 Long Colt is based on original Dragoon conversions of the 1870s. Millington replaced the percussion cylinder with one made from 4140 steel which was then roll engraved in the original 1847 pattern. The conversion itself is of the Richards-Mason type with breechring and ejector. The period-style holster rig, based on the Gus McCrea (Robert Duvall) holster from *Lonesome Dove*, was designed and built by Jim Barnard. The bark ivory grips were handcrafted by Dan Chesiak.

Another Kirst & Strite special edition, this 1849 cartridge conversion begins with either an A. Uberti Baby Dragoon or Wells Fargo model. During the conversion process all the internal parts of the gun are re-heat treated, and the firearm converted to .32 S&W with a Kirst 1849 converter. The entire gun is then proofed and refinished using traditional 19th century techniques. This includes a bone case colored frame and hammer, a charcoal blued barrel and cylinder, and niter blued trigger, screws, and wedge.

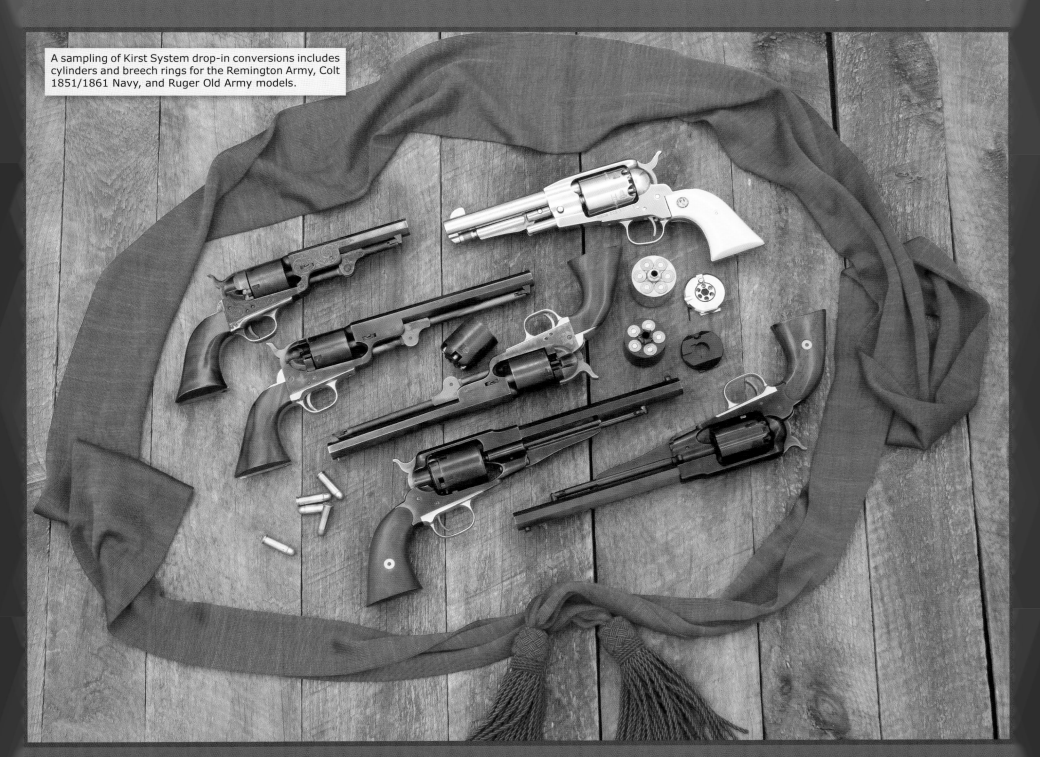

Black Powder Revolvers - Reproductions & Replicas

A sampling of Kirst System drop-in conversions includes cylinders and breech rings for the Remington Army, Colt 1851/1861 Navy, and Ruger Old Army models.

Custom Kirst conversions with antique finish by ArmSport LLC on Uberti 1851 Navy revolvers done without loading gates or ejectors. Though looking similar to a Richards-Mason conversion, the short-barreled Navy can still be switched back to cap and ball in a few minutes. The custom Hickok holster rig was made by Jim Barnard, Navy 5 inch engraving by Conrad Anderson.

The best of both worlds, a completed Kirst conversion allows for the loading and ejection of cartridges without removing the cylinder and breech ring. The conversion is easily done on either a Pietta or Uberti steel frame Remington Army. The kit comes with cylinder, breech ring, and new cylinder pin with attached ejector rod. Cutting a loading channel in the recoil shield is the hardest part of the job. All three versions of Kirst conversion rings are shown; with loading gate (inset on revolver), with loading cutout but no gate, and standard solid ring.

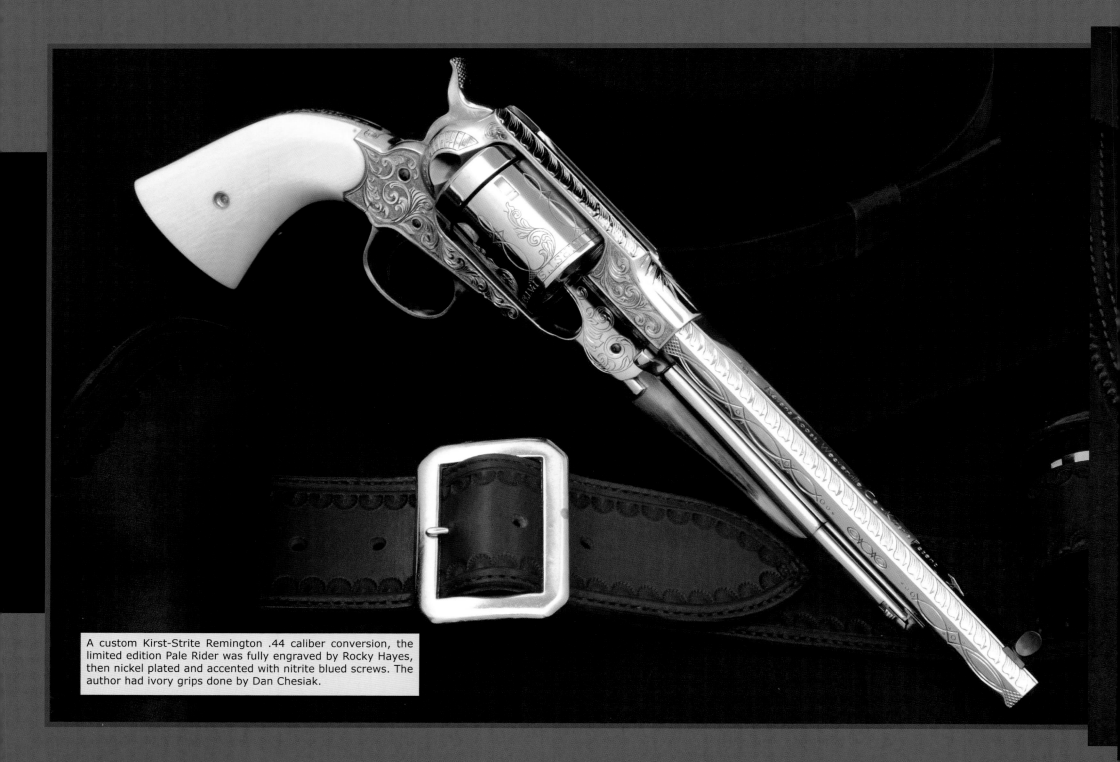

A custom Kirst-Strite Remington .44 caliber conversion, the limited edition Pale Rider was fully engraved by Rocky Hayes, then nickel plated and accented with nitrite blued screws. The author had ivory grips done by Dan Chesiak.

CHAPTER EIGHT: Metallic Cartridge Conversions

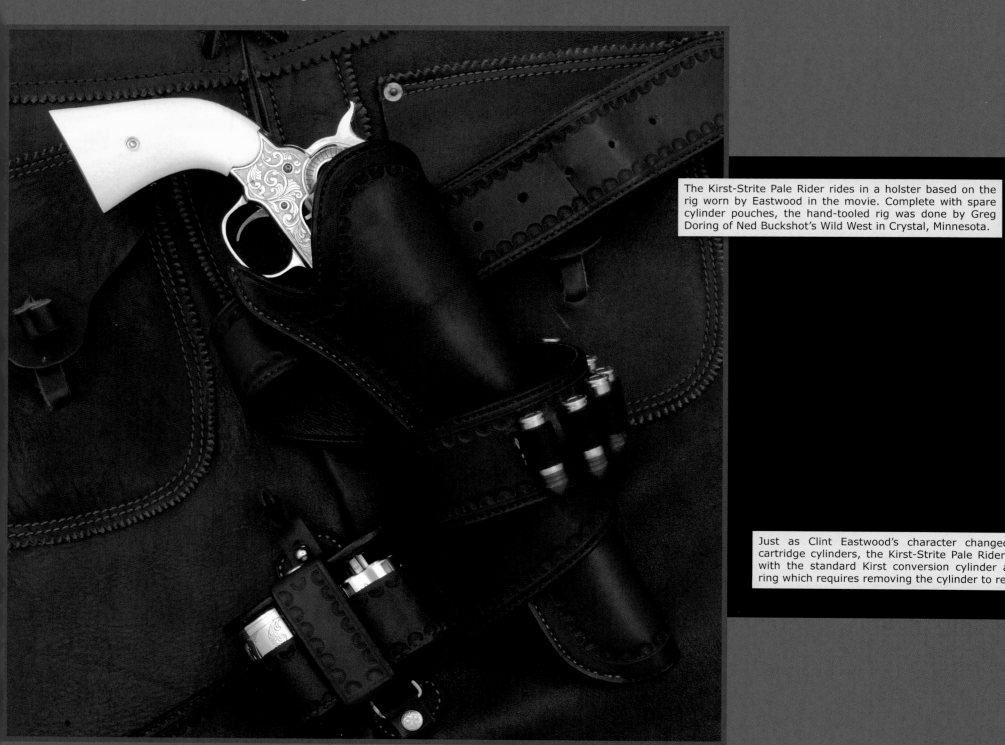

The Kirst-Strite Pale Rider rides in a holster based on the rig worn by Eastwood in the movie. Complete with spare cylinder pouches, the hand-tooled rig was done by Greg Doring of Ned Buckshot's Wild West in Crystal, Minnesota.

Just as Clint Eastwood's character changed loaded cartridge cylinders, the Kirst-Strite Pale Rider is setup with the standard Kirst conversion cylinder and solid ring which requires removing the cylinder to reload.

CHAPTER EIGHT: Metallic Cartridge Conversions

Cimarron F.A Co. and Taylor's & Co. both offer a new Uberti-built Remington conversion chambered in .45 Colt. The gun uses a conversion ring that locks into place using "ears" that come over the sides of the topstrap. The ring also has a loading gate, and the guns come with a period correct manual cartridge ejector.

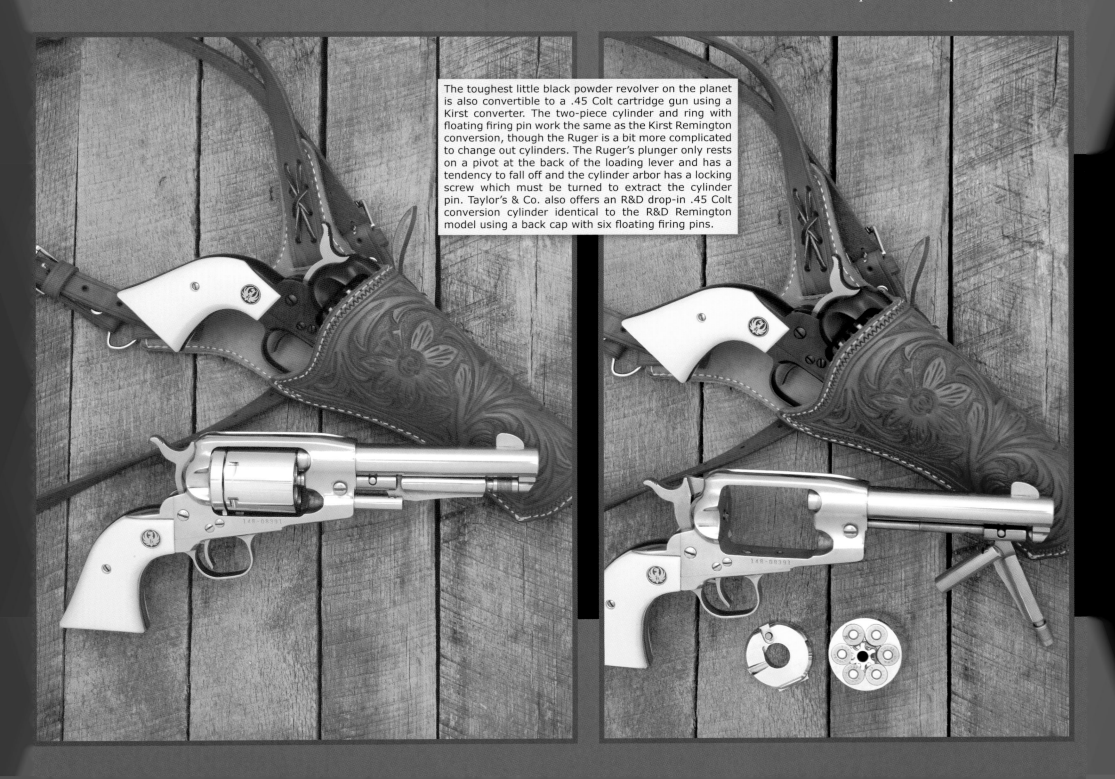

The toughest little black powder revolver on the planet is also convertible to a .45 Colt cartridge gun using a Kirst converter. The two-piece cylinder and ring with floating firing pin work the same as the Kirst Remington conversion, though the Ruger is a bit more complicated to change out cylinders. The Ruger's plunger only rests on a pivot at the back of the loading lever and has a tendency to fall off and the cylinder arbor has a locking screw which must be turned to extract the cylinder pin. Taylor's & Co. also offers an R&D drop-in .45 Colt conversion cylinder identical to the R&D Remington model using a back cap with six floating firing pins.

A second style Remington conversion created by Dave Anderson in the late 1990s was fitted with a loading gate and spring loaded ejector housing. These were built in .38 and .44 caliber.

Trademark Index/Product and Service Sources

John J. Adams, Sr.
Adams & Adams Engraving
P.O. Box 66
Vershire, VT 05079
802-685-0019

America Remembers
10226 Timber Ridge Dr.
Ashland, VA 23005
800-682-2291
www.americaremembers.com

American Historical Foundation, The
10195 Maple Leaf Court
Ashland, VA 23005
804-550-7851
800-368-8080
www.ahfrichmond.com

Conrad Anderson/Rocktree Ranch
9486A Couer d'Alene Rive Rd.
Kingston, ID 83839
208-682-2562
www.rocktreeranch.com

Armi Sport (Armi Chiappa)
via Milano 2
5020 Azzano Mella (BS) Italy
www.armisport.com
info@armisport.com

ArmSport LLC/R. L. Millington
P.O. Box 1308
Platteville, CO 80651
303-810-6411
www.armsportllc.com

Andrew Bourbon Engraving
86 Wm. Maker Way
Brewster, MA 02631
508-896-5159

Black Hills Ammunition
3050 Eglin Street
Rapid City, SD 57703-9574
605-348-5150
www.black-hills.com

Cabelas Inc.
One Cabela Dr.
Sidney, NE 69160
800-237-4444
Fax: 800-496-6329
www.cabelas.com

Cherry's Fine Guns
3408-N West Wendover Avenue
Greensboro, NC 27407
336-854-4182
Fax: 336-854-4184
www.cherrys.com
fineguns@cherrys.com

Cimarron F.A. Co.
P.O. Box 906
Fredericksburg, TX 78624
830-997-9090
www.cimarron-firearms.com

Classic Old West Styles (COWS)
1712 Texas Ave.
El Paso, TX 79901
800-595-2697
www.cows.com

Colt's Manufacturing Company, Inc.
P.O. Box 1868
Hartford, CT 06144-1868
860-236-6311
Fax: 860-244-1442
www.coltsmfg.com
Customer Service Department
800-962-COLT
Fax: 860-244-1449

Dixie Gun Works
1412 W. Reelfoot Ave.
Union City, TN 38261
731-885-0700
www.dixiegunworks.com

E.M.F. Company
1900 E. Warner Ave., Suite 1-D
Santa Ana, CA 92705
949-261-6611
Fax: 949-756-0133
www.emf-company.com
sales@emf-company.com

Euroarms Italia S.r.l.
Importer - EuroArms of America
208 East Piccadily Street
Winchester, VA 22604
540-662-1863
Fax: 540-662-4464
Factory
Via Europa174/C
Concesio, Brescia 25062 Italy
www.euroarms.net
mail@euroarms.net

Trademark Index/Product and Service Sources

Feinwerkbau
Importer - Brenzovich Firearms and Training Center
22301 Texas 20
Fort Hancock, TX 79839
877-585-3775
Fax: 877-585-3775
www.brenzovich.com
bftcgoods@aol.com
Factory - Westinger & Altenburger GmbH
Neckarstrasse 43
Oberndorf, Neckar 78727 Germany
www.feinwerkbau.de
info@feinwerkbau.de

Freedom Arms Inc.
314 Highway 239
Freedom, WY 83120
307-883-2468
www.freedomarms.com
freedom@freedomarms.com.

GOEX Black Powder
Black Dawge Ammunition
Pinnacle Powder
P.O. Box 659
Doyline, LA 71023
318-382-9300
www.goexpowder.com

Gun Works Muzzleloading Emporium, Inc.
247 South 2nd Street
Springfield, OR 97477
541-741-4118
Fax: 541-988-1097
www.thegunworks.com
office@thegunworks.com

Hege
Mengener Str. 38
Messkirch D-88605 Germany
011-49-7575-2872
Fax: 011-49-7575-2872
www.waffen-hege.de

I.A.B. srl
Importer – please refer to E.M.F. listing.
Importer – please refer to Dixie Gun Works listing.
Factory – Industria Armi Bresciane
Via Matteotti, 311
Valtrompia (BS) 25063 Italy
www.iabarms.com
info@iabarms.com

IAR, Inc.
33171 Camino Capistrano
San Juan Capistrano, CA 92675
949-443-3647
Toll-free: 877-722-1873
www.iar-arms.com
sales@iar-arms.com

Jack First Inc.
Gun Parts/Accessories/Service
1201 Turbine Dr.
Rapid City, SD 57703
605-343-9544
Fax: 605-343-9420
www.jackfirstgun.com

Jim Chambers Flintlocks, Ltd.
116 Sam's Branch Road
Candler, NC 28715
828-667-8361
Fax: 828-665-0852
www.flintlocks.com
flintlocks@worldnet.att.net

Kirst Company/Kirst Converters
1544 South Oberlin Circle
Minneapolis, MN 55432
763-571-9220
www.riverjunction.com

Lyman Products Corp.
475 Smith Street
Middletown, CT 06457
800-225-9626
Fax: 860-632-1699
www.lymanproducts.com

Jim Lockwood
Legends in Leather
8100 N. Red Oak Rd.
Prescott, AZ 86305
928-717-0643
www.legendsinleather.com

Trademark Index/Product and Service Sources

Mountain State Manufacturing
Rt. 2 Box 154-1
Williamstown, WV 26187
304-375-2680
Fax: 304-375-7842
www.msmfg.com
msm1@msmfg.com

Navy Arms Co.
219 Lawn St.
Martinsburg, VA 25401
304-262-9870
www.navyarms.com

Ned Buckshot's Wild West
3649 Colorado Avenue North
Minneapolis, MN 55422
763-533-8886

North American Arms, Inc.
2150 South 950 East
Provo, UT 84606
801-374-9990
Toll-free: 800-821-5783
Fax: 801-374-9998
www.naaminis.com

North Star West, Inc.
57 Terrace Court
Superior, MT 59872
406-822-8778
www.northstarwest.com
Laffindog@msn.com

Numrich Gun Parts Corp.
Parts supplier only
226 Williams Lane
P.O. Box 299
W. Hurley, NY 12491
866-686-7424
Fax: 877-486-7278
www.e-gunparts.com
info@gunpartscorp.com

Parsley's Civil War Buckles
8261 Peaks Road
Hanover, VA 23069

Pedersoli, David E & C. Snc.
Distributor - please refer to Cherry's listing.
Distributor - please refer to Dixie Gun Works listing.
Importer - please refer to Cabela's listing.
Importer - please refer to Cimarron, F.A. & Co listing.
Importer - please refer to Navy Arms listing.
Service & Repair – please refer to VTI Gun Parts listing.
Factory
Via Artigiani 57
Gardone V.T. (BS) I-25063 Italy
www.davide-pedersoli.com

Pietta, F.A.P. F.LLI e. C. Snc.
Importer - please refer to Taylor's listing.
Importer - please refer to Dixie Gun Works listing.
Importer - please refer to Cabela's listing.
Importer - please refer to Cimarron, F.A. & Co listing.
Importer - please refer to E.M.F. Co. Inc. listing.
Importer - please refer to Navy Arms listing.
Importer – please refer to IAR listing.
Service & Repair – please refer to VTI Gun Parts listing.
Factory
Via Mandolossa, 102
Gussago (Brescia) 1-25064 Italy
www.pietta.it

Pyrodex Black Powder Substitute
Hodgdon Powder Co.
6231 Robinson
Shawnee Mission, Kansas 66202
913-362-9455
www.hodgdon.com

R&D/Kenny Howell
5728 E. County Rd. X
Beloit, WI 53511
608-921-2594

Remington Arms Co.
870 Remington Drive, P.O. Box 700
Madison, NC 27025
800-243-9700
www.remington.com

Restoration Firearms
6610 Folsom-Auburn Rd., Ste. 5
Folsom, CA 95630
916-791-0596
www.restorationfirearms.com
info@RestorationFirearms.com

Trademark Index/Product and Service Sources

Rifle Shoppe Inc.
18420 E. Hefner Road
Jones, OK 73049
405-396-2583
Fax: 405-396-8450
www.therifleshoppe.com

River Junction Trade Co.
312 Main Street
McGregor IA 52157
563-873-2387
www.riverjunction.com

Stoeger Industries
17601 Indian Head Hwy.
Accokeek, MD 20607-2501
301-283-6981
Fax: 301-283-6988
www.stoegerindustries.com

Stone Mountain Arms, Inc.
5988 Peachtree Corners East
Norcross, GA 30071
770-449-4687
Fax: 770-242-8546

Sturm, Ruger & Co. Inc.
Headquarters
1 Lacy Place
Southport, CT 06490
203-259-7843
www.ruger.com
www.ruger-firearms.com

Service Center for Revolvers, Long Guns & Ruger Date of Manufacture
411 Sunapee Street
Newport, NH 03773
603-865-2442
Fax: 603-863-6165

Taylor's & Co.
304 Lenoir Dr.
Winchester, VA 22603
800-655-5814
www.taylorsfirearms.com

Ten-X Ammunition
5650 Arrow Highway
Montclair, CA 91763
909-605-1617
www.TenXAmmo.com

Thompson/Center Arms Co., Inc.
400 North Main St.
Rochester, NH 03867
603-330-5659
www.tcarms.com

Traditions Performance Firearms
1375 Boston Post Road
P.O. Box 776
Old Saybrook, CT 06475
860-388-4656
Fax: 860-388-4657
www.traditionsfirearms.com
info@traditionsfirearms.com

Trailrider Products/Jim Barnard
P.O. Box 2284
Littleton, CO 80161
303-791-6068
www.gunfighter.com/trailrider

Uberti, A. & C., S.r.l.
Importer - please refer Stoeger Industries listing.
Importer - please refer to Taylor's listing.
Importer - please refer to Dixie Gun Works listing.
Importer - please refer to Cabela's listing.
Importer - please refer to Cimarron, F.A. & Co listing.
Importer - please refer to E.M.F. Co. Inc. listing.
Importer - please refer to Navy Arms listing.
Service & Repair - please refer to VTI Gun Parts listing.
Factory - A. Uberti & C., S.r.l.
Via Artigiani 1
I-25063 Gardone, VT (BS) Italy
www.ubertireplicas.it
info@ubertireplicas.it

United States Fire Arms Mfg. Co.
445 – 453 Ledyard St.
Hartford, CT 06144
860-292-7441
www.usfirarms.com

VTI Gun Parts
P.O. Box 509
Lakeville, CT 06039
860-435-8068
Fax: 860-435-8146
www.vtigunparts.com

Museums of Interest

FIREARMS, WESTERN MEMORIABILIA AND RELATED ARTIFACTS

National Firearms Museum (NFM)
National Rifle Association
11250 Waples Mill Road
Fairfax, VA 22030
www.nrahq.org/museum
(Check website for special exhibits)
This is easily the best firearms museum east of the Mississippi, and if you are an NRA member, please take the time and stop by this well-appointed museum in the Washington, D.C., area - you won't be disappointed. Blue Book Publications, Inc. is proud to be a NRA Foundation supporter of the NFM, and has contributed over $79,000 to date for future museum acquisitions.

Buffalo Bill Historical Center/Cody Firearms Museum
720 Sheridan Ave.
Cody, WY 82414
www.bbhc.org
(Check website for special exhibits)
The BBHC is actually five museums under one large roof - the Cody Firearms Museum, the Plains Indian Museum, Whitney Gallery of Western Art, the Buffalo Bill Museum, and the Draper Museum of Natural History. The BBHC is by far one of the best places on earth to learn about the American West, the Great Plains and early American history. Summer is the busiest time, so check their website for special events and exhibits. Please allow at least two days to take in everything this complex offers, or you will be making a mistake.

Blue Book Publications, Inc. is also pleased to be a One of 1,000 Society sponsor of The Buffalo Bill Historical Center.

Autry National Center of the American West
4700 Western Heritage Way
Los Angeles, CA 90027-1462
www.autrynationalcenter.org
The Autry National Center of the American West is the west coast's premier museum for firearms and western memorabilia. Contains a wide variety of firearms in unique settings. The well landscaped, spacious grounds of Griffith Park are as attractive as the guns and other displays inside this up-to-date museum. Includes the Southwest Museum of the American Indian, the Museum of the American West, and the Institute for the Study of the American West.

The National Cowboy & Western Heritage Museum
1700 NE 63rd Street
Oklahoma City, OK 73111
www.nationalcowboymuseum.org
The only museum in America specializing in cowboy artifacts and memorabilia, including a good selection of original guns used in the west, in addition to modern day reproductions and commemoratives. Original artwork, important artifacts, and many other exhibits make touring this museum mandatory if near Oklahoma City.

J.M. Davis Arms and Historical Museum
333 Lynn Riggs Blvd. (U.S. Route 66)
Claremore, OK 74018
www.state.ok.us
After touring this museum, you'll understand that J.M. Davis never probably turned down a gun for sale! Previously housed in the J.M. Davis Hotel in Claremore, this museum now has its own building, and needs every square inch of space to display the thousands of guns inside. A little bit of everything is represented, including many Colts and Winchesters, although condition on many specimens is below average.

Durham Western Heritage Museum
80 S. 10th St.
Omaha, NE 68108
www.dwhm.org
The Durham Western Heritage Museum provides countless opportunities to discover Omaha's history. After stepping back in time with the life-like sculptures and beautifully restored architecture of the Main Waiting Room, visitors can explore the permanent galleries, including American Indian artifacts, rare coins and documents from the Old West, and detailed histories of the Omaha area.

Frazier Arms Museum
829 W. Main St.
Louisville, KY 48202
www.fraziermuseum.org
One of the more recent museums, don't underestimate Frazier's three stories of displays, with the famous Tower of London exhibits on the third floor, depicting the history of firearms and armor in elaborate dioramas. Additionally, the two floors of American firearms and accessories are well represented in all categories, including pistols, rifles, and shotguns. Many historically significant firearms and memorabilia are displayed, including Teddy Roosevelt's H&H double rifle, Geronimo's bow & arrows, and some guns of Buffalo Bill's.

Museums of Interest

Museum of Connecticut History
231 Capitol Ave.
Hartford, CT 06106
www.museumofcthistory.org

This museum houses the Colt's Patent Firearms Manufacturing Company Factory Collection, donated in 1957. The collection constitutes one of the finest assemblages of early Colt prototypes, factory models and experimental firearms in the world. The collection also includes Colt-made Gatling guns, shotguns and automatic weapons. In 1995 the original "Rampant Colt" statue that had adorned the Hartford Colt factory was acquired by the museum. The Colt Firearms Collection, coupled with historic photographs and other related materials, is a "must-see" for both firearms enthusiasts and students of American history.

Ogden Union Station
2501 Wall Ave.
Ogden, UT 84401
www.theunionstation.org

This complex has five separate museums - the Browning Firearms Museum, the Browning-Kimball Car Museum, Eccles Rail Center, Union Station Natural History Museum and the Utah State Railroad Museum. Additionally, there are two art galleries. The Browning Firearms Museum celebrates the genius of John Browning, inventor of many legendary military and sporting firearms, and many of the prototypes were built in Ogden by Browning himself.

MUSEUMS OF THE CIVIL WAR

National Civil War Museum
One Lincoln Circle at Reservoir Park
Harrisburg, PA 17105-1861
www.nationalcivilwarmuseum.org

Because The National Civil War Museum's mission encompasses the period from 1850 through 1876, its collections vary widely in scope and years of manufacture. For the pre-War period, collections include artifacts that reflect on the nature of sectional controversies and in particular slavery. The military artifacts encompass all aspects of soldiers' experiences: from the personal equipage and weaponry of the War, to wounds, disease, prisoner-of-war experiences, and the emotional drain of the conflict. Whenever possible, emphasis has been placed on obtaining artifacts that are identified to specific combatants of the War, and according to availability, the prominent personalities of the war. Post-War artifacts primarily reflect the impact of the War on western expansion.

American Civil War Museum
297 Steinwehr Ave.
Gettysburg, PA
www.gettysburgmuseum.com

If you could go to one place to learn about the most important battle of the Civil War, Gettysburg is the only choice. The entire city has now been built up to showcase its important history, and how it changed the course of the war. The city features a multitude of tours, shops, reenactments and important museums, including the American Civil War Museum. This museum presents the entire story of the Civil War era and the Battle of Gettysburg with remarkable realism. Learn the causes, effects and significant personalities that shaped the Civil War and as a result, ultimately, American history.

American Civil War Center
490 Tredegar Street
Richmond, VA 23219
www.tredegar.org

The Center is located on a beautiful eight-acre National Historic Landmark site on the James River. Richmond's new Canal Walk fronts the river, and a pedestrian bridge gives visitors access to Belle Isle, a park formerly a Civil War prison camp for captured Union soldiers. Five surviving buildings illustrate the ironworks era and the National Park Service operates the Richmond Civil War Visitor Center, with an outstanding collection of Confederate guns and related artifacts. It's a great place to learn about the Civil War—its causes, its course, and its legacies. It is a place where the people who decided America's future will tell their stories.

Index

Page numbers in italics refer to pages with photos or illustrations.

A

Adams, John J. Sr.............2, 9, 51, 77, 83, 84, 85, 88, *219*

Adams, John J. Jr........................9, 83

Addams, Charles 110

Adler, Dennis 2, *4*, *7*, *8*, *9*, *46*, *71*, *111*, *122*, *123*, *126*, *128*, *130*, *131*, *139*, *150*, *184*, *187*, 188, *194*, *198*, *221*, *223*, *224*, 228

America Remembers.............1, *51*, *79*, 83, *86*, *90*, *92*, *96*, *102*, 104, *106*, *108*, *111*, *114*, *115*

American Frontier Firearms (AFF) ...*190*, *192*, *193*, *194*, *195*, *196*, 199, 200, *202*, 211, *212*, *213*

American Historical Foundation *92*, *112*, *113*, *116*, *117*, *118*, *119*, *120*

American Master Engravers 83

Anderson, Conrad......2, 9, *106*, *216*, *220*, *232*

Anderson, Dave............... *194*, *197*, 199, 200, *204*, *205*, *206*, *207*, *208*, *209*, *210*, *211*, *240*

Anderson, Hunter Scott *178*, *182*, *183*, *184*, *185*

Anderson, Cpl. Mark *187*

Armi Chiappa 4, *157*

Armi San Marco..........................7, 13

Armi San Paolo 7, *153*

ArmSport LLC .... *see Millington, Robert*

Art & Antiques magazine 89

Auto Ordinance 53

Autry, Gene..................................81

Autry Museum 9, 91, 93, 95, 101

B

Baseke, George 188, *189*

Bass Pro..61, 148

Beauregard, Pierre-Gustave Toutant......43, 125

Bedford, PA..............................160, 166, 170, 172, 175, 186, 187

Bennett, E.A 39

Beretta..................................4, 7, 144, 145

Bianchi.. *94*, *95*

Black Hills..................................199, *209*

Blue and Gray 17, *78*

Bluing 65, 154, 155, 156

Bonemeal 152

Book of Colt Firearms, The (Wilson).............59

Bourbon, Andrew.............2, 8, 51, 80, 85, 89

Buffoli, Angelo 5, 13

Burr, David 43, 134, 143, 160

Burns, Ken 29

Burton, James Henry....................43

Butterfield, E.B 39

C

Cabela's 9, *46*, 61, *66*, *69*, 71, 147, 148

Cadillac..13

Campeche, Battle of 15

Cartridge Conversions.... *see Metallic Cartridge Conversions*

Case hardening......43, 65, *152*, 154, *155*, 156

Chamberlain, Col. Joshua Lawrence............16

Chaney, William R......................79

Charcoal......................................152

Charging.........................*see Loading*

Charleston harbor 43

Chesiak, Dan *220*, *229*, *234*, *235*

Chinetti, Luigi 143

Churchill, Winston G 85

Cimarron, F.A. Co *40*, 61, *62*, 65, 66, 67, 68, 69, 147, 148, 199, *221*, *223*, 238

Civil War..................4, 5, 6, 8, 17, 20, 21, 29, 34, 37, 38, 41, 43, 44, 45, 47, 48, 50, 64, 65, 67, 69, 77, 78, 79, 81, 83, 99, 107, 109, 113, 114, 115, 118, 119, 123, 124, 125, 147, 148, 153, 157, 158, 159, 160, 161, 175, 177, 178, 179, 180, 181, 183, 220

Civil War Belt Buckles............................. *224*

Cleaning........... *132*, *133*, *134*, *135*, *136*, *137*

Clothing, Civil War..........................*158*, *159*, *160*, *161*, *162*, *163*, *164*, *165*, *166*, *167*, *168*, *169*, *170*, *171*, *172*, *173*, *175*, *186*, *187*

Clothing, Western/Cowboy..............*178*, *179*, *180*, *181*, *182*, *183*, *184*, *185*, *187*, *188*, *189*

Cody, Wm............................. .1, 5, *49*, *74*, 79

Colburn, David G........................39

Colt, Colonel Samuel 3, 5, 6, 11, 13, 15, 17, 20, 22, 29, 33, 35, 39, 45, 59, 61, 67, 75, 77, 87, 99, 103, 104, 109, 112, 114, 116, 117, 120, 122, 147, 229

Colt Blackpowder Arms Company 2, 9, 10, 15, 17, 18, 19, 29, 33, 34, 35, 77, 83, 83, 84, *151*, *152*, 154, *155*, 156, *157*

Colt Blackpowder production data..15, 17, 19

Index

Colt Blackpowder Reproductions & Replicas.. ....4, 7, 8, 9, 111

Colt Guns

    1st Generation ....11, 13, 15, 17, 20, 21, 22, 23, 77, 87, 89, 125, 191, *196* (Richards conversion)

    2nd Generation

        Model Specifications............................27

        Army 1860 ....*10, 16, 17, 18, 19,* 20, 21, 22, *23,* 27, 29, *33, 37, 40, 64, 65, 69, 71, 73, 74,* 77, *84, 97, 98,* 147

        Cavalry, U.S...............................*83, 97, 98*

        Dragoon

            1st Model *12,* 20, *23, 25,* 27, 29, *85*

            2nd Model *14,* 20, *23, 25,* 27

            3rd Model *14, 20, 23, 25,* 27, *80, 91*

            Baby 1848 ....*26,* 27, *85*

            Dragoon production data 25

        Navy 1851.....................*16, 17, 19,* 20, 22, *26,* 27, 29, *40, 50,* 65, *78, 86, 101,* 147

        Navy 1851, Reb Short *49, 70, 72,* 147

        1851 U.S. Grant Comm 78

        1851 R. E. Lee Comm..........................78

        Navy 1861 ....*18,* 22, 25, 27

            Navy 1861 production data..............22

        Pocket Navy 1862...............*21,* 22, 27, 33, *84, 120,* 181

            Pocket Navy 182 production data......22

        Pocket Police 1862..............*21,* 22, 27, 29

        Walker 1847 ....7, *12,* 13, 15, *23,* 27, 29, 30, *83, 87, 90, 93*

    Heritage 1847.....................15, 27, *82,* 83

        Accessories...................*35, 77, 83, 85, 103*

    3rd Generation

        Model specifications 30

        Army 1858 .... *see Remington Revolvers*

        Army 1860 .... 2, 31, *33, 37, 40, 64,* 65, *67, 69, 71, 73, 74, 102, 106, 107, 112, 113, 117, 174, 177, 196,* 211, 214, 220, 222, 223

        1860 Gold U.S. Cavalry 31, 33, *92, 94,* 95

        1860 Tiffany *10,* 31, 33, *79, 84, 89*

        1871 Open Top Tiffany *190, 192*

        Dragoon

            Little Big Horn (Tiffany)...................*81*

            1st Model 1848 .... 30, *34, 36, 64, 110, 114, 177,* 229

            2nd Model 1850 .. 30, *64,* 65, *118, 177*

            3rd Model *28,* 30, *64, 80, 91, 118, 174, 177*

                Production data.............................20

            3rd Model 1851 Cochise...............*10,* 30, 33, *82*

        Baby 33, 68, 112, 174, 228, 229, 230

            Baby 1848................25, 30, *200,* 229

        California Commemorative 95

        Whitneyville-Hartford 1848 ....23, 29, 30, *32, 33, 111*

        Marine 30, *94*

        Navy 1851 ....2, 31, *32, 33, 36, 40, 64,* 65, *68, 69, 99, 101, 103, 104, 105, 106, 109, 113, 114, 117, 119, 120, 123, 149,*160, *174, 177,* 182, *194,* 209, 212, 213, 224, 231

        Navy 1861 .... *24,* 31, *32, 33, 64,* 65, *85,* 130, *137, 174, 193, 194, 197,* 202, 205, 210, 226, 231

        Navy 1861 Prototype *24,* 25

        Paterson 1842 ... 3, *10,* 33, *76, 96, 107, 117* (see listings under Paterson)

        Pocket 1849 .... 29, 31, *32, 33, 36,* 230

        Pocket Navy 1862.......................29, 31, *32, 33, 64, 66, 70, 174, 195,* 218

        Pocket Police 1862... 31, *32, 33, 64, 66, 67, 116, 200,* 214

        Trapper 1862 ....*29,* 31, *32, 33, 36,* 61

        Walker 1847 ....7, 30, *32, 62, 63, 64, 66, 86, 87, 90, 108, 116,* 123, *174, 177, 189, 228, 229*

        Accessories 35

Confederate States of America...............41, 51

Conversions ....*see Metallic Cartridge Conversions*

Copyright, Credits, & Cover Description......2

Classic Old West Styles (COWS) *177, 184, 188, 194*

Colt Heritage, The (Wilson) 15, 82

Colt Single Action – From Patersons to Peacemakers...219

Costner, Kevin 29, 144

Cowboy Action Shooting 7, 29, 159, 178, 179, 180, 181, 182, 183, 185

Custer, General George Armstrong.............33, 85, 161, *187*

Index

Cyanide 152

D

Daly, Bruce *32, 190, 192, 210*
Dance, J.H 3,
 41, 43, 48, *49, 50, 51,* 147, 148
Davis, Jefferson 2, 41, 99, 104, 120
Dick Armagost Saddler Shop *176, 209*
Disassembly *138, 139,*
Dixie Gun Works 61,
 130, 144, 147, 148, 153, *187,* 199
Dodds, Isaac 39
Dove, Howard 85, 95
Drop-in Conversions *228, 229*
 Kirst Konverter 201, 203, *226, 227, 230, 231, 232, 233, 234, 235, 236, 237*
Duvall, Robert 229

E

EMF 148
Earp, Wyatt 178
Eastwood, Clint 6, 8, 9, 45, 143, 144, 236
Edwards, David 39
El Paso Saddlery *12*
Engraving....*see individual images, engravers are listed alphabetically by name and/or company*
Euroarms 4, 8, 48, *50,* 153
Everhart, Duncan *77*

F

Faries, Samuel 39
Feinwerkbau 157
Filback, Tom *179, 180, 182, 184*
Finishes and Fit 63, 69, *230*
 (*see Bluing, Case hardening, and Patina listings*)
Fjestad, S.P 2, *4, 7, 8, 142,* 143
Fonda, Henry 178
Ford, Gerald 85
Ford, Henry 13
Foreword 5-7
Forgett, Val Sr 5, 6, *7,* 8, 9, 15, 26, 59, 143, 144, 160
Forgett, Val Jr 5, *7,* 9
Francolini, Leonard 85
Freeman, Austin 47
French Lefaucheux 41, 195

G

Geronimo 43, *51*
Gettysburg 160, 187
Gettysburg, Battle of 107
Girard, Charles Frederic 43
Girard & Cie 43
Goex FFFg 56, 127, *133,* 140, 199
Grant, Ulysses S 17, 26, 27, 33, *78,* 161
Gregorelli 147
Griswold, Samuel 3, 39, 41

Griswold and Gunnison 3, 41,
 48, *49,* 148, 153, *174*
Gunnison, A 41

H

Hable, Bob 61
Hartford, CT 15
Hartley, Marcellus 41, 195
Harvey, Mike 9, 69
Hatco *54*
Haviland, F.B 39
Hayes, Rocky *234, 235*
Hays, Col. Jack..................................... 114
Hickok, Wild Bill.....3, 15, 64, 113, *119,* 123, 178, 180
Helfricht, Cuno A 84, 87, *88*
Heard, Dow 218
Hoard, C.B 47
Holland, Dennis................................... *220*
Holland & Holland................................ 152
Holsters..... *12,* 19, 22, 34, 36, 55, 60, 72, 87, *96, 123, 177, 178, 190, 191, 192, 193, 194, 198, 200, 204, 208, 210, 213, 215, 219, 224, 226, 228, 229, 232, 234, 236, 238, 239*
Hood, Graham 75, 77
Howell, Kenny 2, 199, 203, *224, 225*
Hunt, K.C.................................... 85, *91,* 101
Hurst, Ken 85

Index

I

Imperato, Anthony.........8, 9, 29, 81, 152, *157*
Imperato, Louis...................9, 26, 29, 81
Introduction...4
Iver Johnson Arms..............................26

J

Jabar, Kareem Abdul........................144
Jackson, William 109
Jacquith, Elijah....................................39
Johnson, Lyndon B 85
Jones, Com. John Paul 114, *115*

K

Kennedy, John F..................................85
Kilmer, Val 144, 184
Kirkland, Turner................................144
Kirst, Walt....2, 201, *226, 227, 228, 230, 231, 232, 233, 234, 236, 239*
Kirst & Strite ....see Strite listing
Kittredge & Co., B.............................220
Klay, Frank...................................*77, 103*
Krider, John..43

L

Lantuch, Paul......................................59
Lee, Robert E 17, 26, 27, 33, *78*, 109, 113, 161

Legends in Leather (Jim Lockwood)... 9, *60, 219*
Leland, Henry Martyn 13
LeMat revolver 2, 3, *38, 42*, 43, 48, *49*, 52, *92*, 125, 147, 148, 149, *176*
LeMat, Dr. Jean Alexandre Francois 42, 43, 45, 48, 59
Lemat The Man, The Gun (Forgett & Serpette) 42, 59
LePretre, Justine Sophie......................43
Lewis, Ben...*186*
Lincoln, President Abraham 41, 195
Lindsay, Merrill.................................*110*
Lione, Sergio...............................143, 144
Little Round Top, Battle of................125
Loading....*125, 126, 127, 128, 129, 130, 131, 140*, 141
Longley, Bill.......................................43

M

Macon, GA..43
Maintenance .... see Loading, Disassembly and Cleaning
Martin, Greg.......................................89
Masterson, Bat..................................178
McCulloch, Ben.................................114
Metallic Cartridge Conversions *190-240*
Also see individual listings by manufacturer/customizer
Metallurgy..87
Mexican War......................................15
Miller, J..39

Millington, Robert (ArmSport LLC)............2, 9, *60*, 69, *96*, 198, 201, *214, 215, 216*, 218, *219, 220, 222, 228, 229, 232*
Mitchell, Jeff................................122, *123*
Mitchum, Robert 144
Ming, Dennis "China Camp"... *181, 182, 184*
Mix, Tom...178
Mosby, Col. John Shingleton.............113, 123
Movies
  Dances With Wolves...................29, 144
  Gettysburg..................................29, 159
  Gods & Generals........................29, 159
  Good, Bad, & The Ugly, The.........8, 65, 143
  Outlaw Josey Wales, The 8, 143
  Quick & The Dead, The.........................199
  Silverado...144
  Tombstone..........................144, 159, 184
  Unforgiven............................... *45*, 143
  Wyatt Earp 144
  Young Guns 144
Murphy, Dr. Joseph A 9, 89

N

Navy 1851, buffalo...............................*60*
Navy Arms 9, 26, *42*, 61, *72*, 144, 147, 160
Ned Buckshots Wild West........ *224, 236, 237*
Nimschke, Louis 33, 85, 88, 89
Nutting, Mighill..................................39

Index

O
Ormsby, Waterman Lilly 33, *84*

P
Palmetto...148, 153
Patents 11
Patent Arms Manufacturing Company.......13, 30, 39
Paterson Colt Pistol Variations......................107
Paterson models *10*, 11, 13, 25, 30, *33*, 39, 60, *61*, *62*, *63*, *64*, *76*, 77, 83, *96*, *107*, *117*, 125, 147
Patina..65
Pearson, John 11
Pedersoli, Davide..................................7, 157
Pettingill revolver 47
Phillips, Philip R.......................................107
Pietta, Alberto *148*, *150*
Pietta, Alessandro *148*, *150*
Pietta, Giuseppe .... *148* , *150*
Pietta, Fratelli...2, 4, 8, 9, *38*, *44*, *46*, *47*, *48*, *49*, *50*, *60*, *61*, *63*, 65, *67*, *69*, *70*, *71*, *73*, 143, *147*, *148*, *149*, *150*, 157, *174*, *198*, 233
Pinfire ignition.................................191, 195
Plains Indian Wars 21
Preface & Acknowledgements 8-9
Purdey...152
Pyrodex P.................. 56, *124*, 127, 133, *135*, 139, *140*, 141

R
R&D Gun Shop....................... *224*, *228*, *239*
Reagan, Ronald...85
Reenactments, Civil War...........*158*, *159*, *160*, *161*, *162*, *163*, *164*, *165*, *166*, *167*, *168*, *169*, *170*, *171*, *172*, *173*, *174*, *175*, *186*, *187*
Reenactments, Cowboy *178*, *179*, *180*, *181*, *182*, *183*, *184*, *185*, *188*, *189*
Remington, Eliphalet 5, 39
Remington Revolvers 1, 2, 3, *40*, *41*, 45, *46*, *47*, *48*, *49*, 52, 65, *69*, 71, 86, *106*, 125, 129, 148, *174*, *179*, *194*, *231*, *233*, *234*, *240*
Richards Conversions............... *193*, *194*, *196*, *197*, 199, 200, 201, *202*, *205*, *210*, *211*, *212*, *213*, *216*, *217*, *219*, *223*, *226*, *227*, *228*
Richards-Mason Conversions ...*200*, *215*, *222*, *224*, *225*
Richards-Mason195, 197, 199, *200*, *209*, *214*, 213, *214*, *215*, *222*, *224*, *225*, *228*, *229*
Richmond, VA 41
Rogers & Spencer........3, 47, 48, *50*, 153, 157
Root pistol *153*, *187*
Ruger & His Guns (Wilson) 5, 55, 59
Ruger, Bill Jr 9, 52, 53, 54
Ruger Collectors Association, ...57, *58*, *59*
Ruger Old Army 5, *52*, *53*, *54*, 55, *56*, *57*, *58*, *59*, *124*, 125, *126*, 127, *128*, 129, *131*, *132*, *134*, *136*, *139*, 180, *231*
    Production data 56, 57, 58
Ruger, William B. Sr *5*, 6, 52, 53, 56, 145
Russell, Bob *188*, *189*
Russell, Dennis..................................9, 19, *78*

S
Safety ....*see Loading, Disassembly & Cleaning*
Samuel Colt Presents (Wilson) 87
Schyler, Hartly & Graham ... 41, *84*, 195, 220
Sefreid, Harry..55, 56
Selleck, Tom 6, 144, 203
Serezzo 143
Sheen, Charlie..144
Shockey, Chris...*186*
Shooting...*see Loading, Disassembly & Cleaning*
SHOT Show 6, 69
Single Action Shooting Society (SASS)...8, *53*, 83, *179*, 181
Smith & Wesson 191
Smithsonian Institution...............................43
Spangenberger, Phil...................................144
Spiller & Burr 3, 43, 48, *49*, 147, 148, *174*
Spiller, Edward 43
Spring, George 85
Starpacker Badges....................................*222*
Starr, Ebenezer Townsend..............45, 47, 201
Starr Percussion Pistols.... 2, 3, *44*, 45, 47, 48, 125, 143, 147, *148*, *149*, *198*, 199, 201
    Production data 199, 201
Steel Canvas (Wilson) 81
Stetson *94*
Stewart, James...178
Stone, Sharon..144
Strichman, George A.....*5*, 9, 81, *91*, *93*, 95, *101*

Index

Strite (Kirst & Strite).... *230, 234, 235, 236, 237*
Stuart, Gen. Jeb......................43, *92*
Sturm, Ruger & Company...............9, 53, 56
Sumter, Fort...........................43, 125

T

Table of Contents..........................3
Taylor's & Company 7, 9, *40*, 61, *66, 68, 71*, 130, 141, 147, 148, 199, 203, *228, 229, 238, 239*
Television shows/movies
    Deadwood...............................159
    Dr. Quinn, Medicine Woman 199
    Lonesome Dove.....................144, 229
    Story of the Gun.....................7, 144
    Tales of the Gun..........................7
Ten-X..................................199
Texas Jack's.......9, *40, 62, 63, 64, 187, 188, 189*
Texas Paterson Pistol *62*
Texas Rangers......................... 13, 25, 33, *114*
Texas, Republic of 17
Thirion, Denise...........................85
Thornber, Tom............................13
Tiffany, Charles Lewis 79
Tiffany & Company.........................2, *10*, 77, 79, 81, 83, *84*, 85, *89, 190*
Title page 1
Topinka, Joe............................*187*
Torme, Mel..............................144

Tracy, Hugh *32, 85*
Trailrider Products.............*228, 229, 232*
Trademark Index 241-246
Triggs, James M..........................59

U

U.S. Fire Arms Mfg....................*49*
Uberti, Aldo.... 6, 7, 8, 15, *66, 116, 142*, 143, 144, 145, 147
Uberti, Maria..........................6, 145
Uberti & Company....................4, 7, 8, *40, 41*, 48, *61, 62, 63, 64, 65, 66, 67, 68, 69, 70*, 73, *104, 107, 142*, 143, *144, 145, 146*, 157, *189*, 199, *221, 223, 232, 233, 238*
United States Historical Society............. 2, *74, 92, 99, 107, 109, 110*
U.S. International Muzzle Loading Team ..133

V

VTI Gun Parts 60

W

Wah Maker Frontier Clothing..........*178, 179, 180, 181, 183, 184, 185, 187, 188, 189*
Walker......*see individual listings by manufacturer*
Walker, Capt. Samuel....................15, *86, 93*
Watts, Tom..............................89
Wayne, John............................178
Webb, Suzanne...........................69
Wells Fargo..........................68, *112*

Western Expansion of the 1870s 125
White, Alvin A........8, 80, 83, 85, *91, 93*, 101
Whitney, Eli............................5, 15
Whittier, Otis W 39
Wilson, R.L......................... *4, 5, 6, 7, 8*, 13, 20, 21, 55, 59, 75, 77, 79, 81, *82*, 83, 85, 87, 89, *107, 110*, 143, 144
Wonder Wads.. *124, 126*, 127, *128*, 129, *130, 131, 140*, 141
Woodson, Jack *74, 92, 106*

Y

Young, Gustave ...80, 85, 87, *88*, 89, *101, 103*

39.95   11/2/12

PUBLIC LIBRARY
800 Middle Country Road
Middle Island, NY 11953
(631) 924-6400
mylpl.net

LIBRARY HOURS

Monday-Friday	9:30 a.m. - 9:00 p.m.
Saturday	9:30 a.m. - 5:00 p.m.
Sunday (Sept-June)	1:00 p.m. - 5:00 p.m.